OFFICIAL SQA PAST PAPERS

WITH ANSWERS

HIGHER

PHYSICS
2006-2009

First exam published in 2006.
Published by Bright Red Publishing Ltd, 6 Stafford Street, Edinburgh EH3 7AU
tel: 0131 220 5804 fax: 0131 220 6710 info@brightredpublishing.co.uk www.brightredpublishing.co.uk

ISBN 978-1-84948-068-0

A CIP Catalogue record for this book is available from the British Library.

Bright Red Publishing is grateful to the copyright holders, as credited on the final page of the book, for permission to use their material. Every effort has been made to trace the copyright holders and to obtain their permission for the use of copyright material. Bright Red Publishing will be happy to receive information allowing us to rectify any error or omission in future editions.

[BLANK PAGE]

X069/301

NATIONAL
QUALIFICATIONS
2006

WEDNESDAY, 17 MAY
1.00 PM – 3.30 PM

PHYSICS
HIGHER

Read Carefully

Reference may be made to the Physics Data Booklet.

1 All questions should be attempted.

Section A (questions 1 to 20)

2 Check that the answer sheet is for Physics Higher (Section A).

3 For this section of the examination you must use an **HB pencil** and, where necessary, an eraser.

4 Check that the answer sheet you have been given has **your name**, **date of birth**, **SCN** (Scottish Candidate Number) and **Centre Name** printed on it.

   Do not change any of these details.

5 If any of this information is wrong, tell the Invigilator immediately.

6 If this information is correct, **print** your name and seat number in the boxes provided.

7 There is **only one correct** answer to each question.

8 Any rough working should be done on the question paper or the rough working sheet, **not** on your answer sheet.

9 At the end of the exam, put the **answer sheet for Section A inside the front cover of your answer book**.

10 Instructions as to how to record your answers to questions 1–20 are given on page three.

Section B (questions 21 to 29)

11 Answer the questions numbered 21 to 29 in the answer book provided.

12 **All answers must be written clearly and legibly in ink**.

13 Fill in the details on the front of the answer book.

14 Enter the question number clearly in the margin of the answer book beside each of your answers to questions 21 to 29.

15 Care should be taken to give an appropriate number of significant figures in the final answers to calculations.

SCOTTISH
QUALIFICATIONS
AUTHORITY

©

DATA SHEET
COMMON PHYSICAL QUANTITIES

Quantity	Symbol	Value	Quantity	Symbol	Value
Speed of light in vacuum	c	$3 \cdot 00 \times 10^{8}$ m s^{-1}	Mass of electron	m_e	$9 \cdot 11 \times 10^{-31}$ kg
Magnitude of the charge on an electron	e	$1 \cdot 60 \times 10^{-19}$ C	Mass of neutron	m_n	$1 \cdot 675 \times 10^{-27}$ kg
Gravitational acceleration on Earth	g	$9 \cdot 8$ m s^{-2}	Mass of proton	m_p	$1 \cdot 673 \times 10^{-27}$ kg
Planck's constant	h	$6 \cdot 63 \times 10^{-34}$ J s			

REFRACTIVE INDICES

The refractive indices refer to sodium light of wavelength 589 nm and to substances at a temperature of 273 K.

Substance	Refractive index	Substance	Refractive index
Diamond	2·42	Water	1·33
Crown glass	1·50	Air	1·00

SPECTRAL LINES

Element	Wavelength/nm	Colour	Element	Wavelength/nm	Colour
Hydrogen	656	Red	Cadmium	644	Red
	486	Blue-green		509	Green
	434	Blue-violet		480	Blue
	410	Violet	Lasers		
	397	Ultraviolet	Element	Wavelength/nm	Colour
	389	Ultraviolet	Carbon dioxide	9550 } 10590	Infrared
Sodium	589	Yellow	Helium-neon	633	Red

PROPERTIES OF SELECTED MATERIALS

Substance	Density/ kg m^{-3}	Melting Point/ K	Boiling Point/ K
Aluminium	$2 \cdot 70 \times 10^{3}$	933	2623
Copper	$8 \cdot 96 \times 10^{3}$	1357	2853
Ice	$9 \cdot 20 \times 10^{2}$	273	. . . .
Sea Water	$1 \cdot 02 \times 10^{3}$	264	377
Water	$1 \cdot 00 \times 10^{3}$	273	373
Air	$1 \cdot 29$	. . . .	. . . .
Hydrogen	$9 \cdot 0 \ \times 10^{-2}$	14	20

The gas densities refer to a temperature of 273 K and a pressure of $1 \cdot 01 \times 10^{5}$ Pa.

SECTION A

For questions 1 to 20 in this section of the paper the answer to each question is either A, B, C, D or E. Decide what your answer is, then, using your pencil, put a horizontal line in the space provided—see the example below.

EXAMPLE

The energy unit measured by the electricity meter in your home is the

    A kilowatt-hour

    B ampere

    C watt

    D coulomb

    E volt.

The correct answer is **A**—kilowatt-hour. The answer **A** has been clearly marked in **pencil** with a horizontal line (see below).

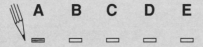

Changing an answer

If you decide to change your answer, carefully erase your first answer and, using your pencil, fill in the answer you want. The answer below has been changed to **E**.

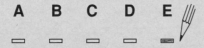

[Turn over

SECTION A

Answer questions 1–20 on the answer sheet.

1. Which of the following contains one scalar quantity and one vector quantity?

   A   acceleration; displacement

   B   kinetic energy; speed

   C   momentum; velocity

   D   potential energy; work

   E   power; weight

2. A golfer strikes a golf ball which then moves off at an angle to the ground. The ball follows the path shown.

   The graphs below show how the horizontal and vertical components of the velocity of the ball vary with time.

   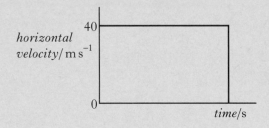

   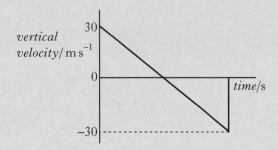

   What is the speed of the ball just before it hits the ground?

   A $10\,\mathrm{m\,s^{-1}}$

   B $30\,\mathrm{m\,s^{-1}}$

   C $40\,\mathrm{m\,s^{-1}}$

   D $50\,\mathrm{m\,s^{-1}}$

   E $70\,\mathrm{m\,s^{-1}}$

3. An object starts from rest and accelerates in a straight line.

   The graph shows how the acceleration of the object varies with time.

   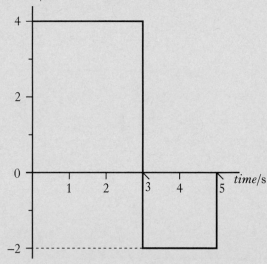

   The object's speed at 5 seconds is

   A $2\,\mathrm{m\,s^{-1}}$

   B $8\,\mathrm{m\,s^{-1}}$

   C $12\,\mathrm{m\,s^{-1}}$

   D $16\,\mathrm{m\,s^{-1}}$

   E $20\,\mathrm{m\,s^{-1}}$.

4. A person stands on bathroom scales in a lift. The scales show a reading greater than the person's weight.

   The lift is moving

   A   upwards at constant velocity

   B   downwards at constant velocity

   C   downwards and accelerating

   D   downwards and decelerating

   E   upwards and decelerating.

5. Two trolleys travel towards each other in a straight line as shown.

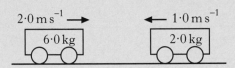

The trolleys collide. After the collision the trolleys move as shown below.

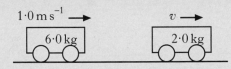

What is the speed v of the 2·0 kg trolley after the collision?

A 1·25 m s^{-1}

B 1·75 m s^{-1}

C 2·0 m s^{-1}

D 4·0 m s^{-1}

E 5·0 m s^{-1}

6. A student carries out an experiment to determine the density of a liquid.
The results are shown.

volume of liquid in beaker $= 2\cdot00 \times 10^{-5}$ m^3
mass of empty beaker $= 3\cdot00 \times 10^{-2}$ kg
mass of filled beaker $= 4\cdot50 \times 10^{-2}$ kg

The density of the liquid is

A $4\cdot44 \times 10^{-4}$ kg m^{-3}

B $1\cdot33 \times 10^{-3}$ kg m^{-3}

C $7\cdot50 \times 10^{2}$ kg m^{-3}

D $2\cdot25 \times 10^{3}$ kg m^{-3}

E $3\cdot75 \times 10^{3}$ kg m^{-3}.

7. Which pair of graphs shows how the pressure produced by a liquid depends on the depth and density of the liquid?

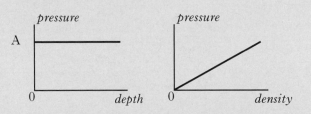

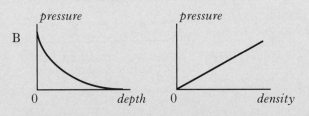

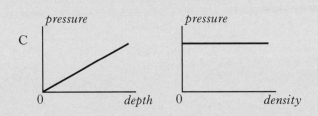

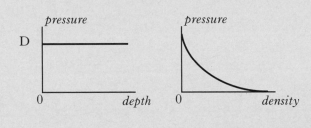

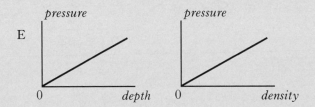

[Turn over

8. Three resistors are connected as shown.

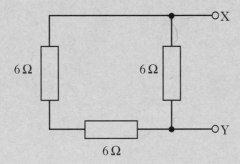

The total resistance between X and Y is

A $2\,\Omega$

B $4\,\Omega$

C $6\,\Omega$

D $9\,\Omega$

E $18\,\Omega$.

9. A battery of e.m.f. 12 V and internal resistance $3\cdot0\,\Omega$ is connected in a circuit as shown.

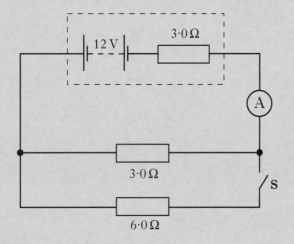

When switch **S** is closed the ammeter reading changes from

A $2\cdot0\,A$ to $1\cdot0\,A$

B $2\cdot0\,A$ to $2\cdot4\,A$

C $2\cdot0\,A$ to $10\,A$

D $4\cdot0\,A$ to $1\cdot3\,A$

E $4\cdot0\,A$ to $6\cdot0\,A$.

10. The circuit diagram shows a balanced Wheatstone bridge.

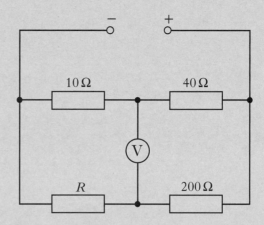

The resistance of resistor R is

A $0\cdot5\,\Omega$

B $2\cdot0\,\Omega$

C $50\,\Omega$

D $100\,\Omega$

E $800\,\Omega$.

11. A student carries out three experiments to investigate the charging of a capacitor using a d.c. supply.

The graphs obtained from the experiments are shown.

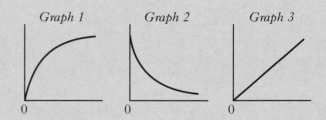

The axes of the graphs have not been labelled.

Which row in the table shows the labels for the axes of the graphs?

	Graph 1	Graph 2	Graph 3
A	voltage and time	current and time	charge and voltage
B	current and time	voltage and time	charge and voltage
C	current and time	charge and voltage	voltage and time
D	charge and voltage	current and time	voltage and time
E	voltage and time	charge and voltage	current and time

12. The following circuit shows a constant voltage a.c. supply connected to a resistor and capacitor in parallel.

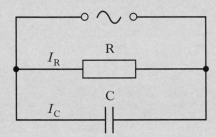

Which pair of graphs shows how the r.m.s. currents I_R and I_C vary as the frequency f of the supply is increased?

A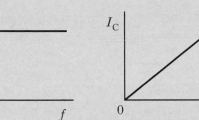

B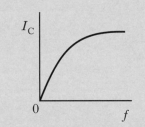

C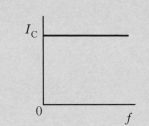

D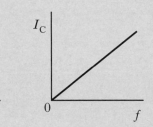

E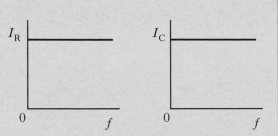

13. An op-amp circuit is set up as shown.

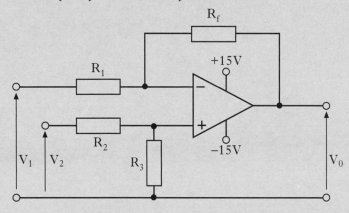

Which of the following statements is/are true?

I   The circuit is connected in the inverting mode.

II The circuit amplifies the difference between V_2 and V_1.

III The maximum possible output voltage $V_0 = V_1 + V_2$.

A   I only

B   II only

C   I and II only

D   II and III only

E   I, II and III

14. A microwave source at point O produces waves of wavelength 28 mm.

A metal reflector is placed as shown.

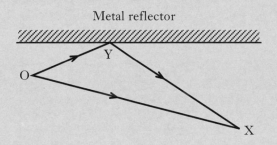

An interference pattern is produced. **Constructive interference** occurs at point X. The distance OX is 400 mm. The total path length OYX is

A   414 mm

B   421 mm

C   442 mm

D   456 mm

E   463 mm.

15. The diagram represents a ray of light passing from air into liquid.

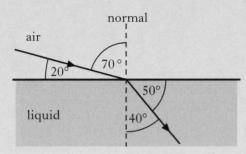

The refractive index of this liquid, relative to air, is

A $\dfrac{\sin 20°}{\sin 40°}$

B $\dfrac{\sin 40°}{\sin 70°}$

C $\dfrac{\sin 50°}{\sin 20°}$

D $\dfrac{\sin 70°}{\sin 40°}$

E $\dfrac{\sin 90°}{\sin 40°}$.

16. Light travels from air into glass.

Which row in the table describes what happens to the speed, frequency and wavelength of the light?

	Speed	Frequency	Wavelength
A	increases	stays constant	increases
B	increases	decreases	stays constant
C	stays constant	decreases	decreases
D	decreases	decreases	stays constant
E	decreases	stays constant	decreases

17. When light of frequency f is shone on to a certain metal, photoelectrons are ejected with a maximum velocity v and kinetic energy E_k.

Light of the same frequency but twice the irradiance is shone on to the same surface.

Which of the following statements is/are correct?

I   Twice as many electrons are ejected per second.

II   The speed of the fastest electron is $2v$.

III   The kinetic energy of the fastest electron is now $2E_k$.

A   I only

B   II only

C   III only

D   I and II only

E   I, II and III

18. The diagram shows some of the energy levels for the hydrogen atom.

E_3 ———————— -1.360×10^{-19} J
E_2 ———————— -2.416×10^{-19} J

E_1 ———————— -5.424×10^{-19} J

E_0 ———————— -21.76×10^{-19} J

The highest frequency of radiation emitted due to a transition between two of these energy levels is

A 1.59×10^{14} Hz

B 2.46×10^{15} Hz

C 3.08×10^{15} Hz

D 1.63×10^{20} Hz

E 2.04×10^{20} Hz.

19. A series of radioactive decays starts from the isotope Uranium 238.

Two alpha particles and two beta particles are emitted during the decays.

Which row in the table gives the mass number and the atomic number of the resulting nucleus?

	Mass number	Atomic number
A	232	88
B	230	86
C	230	90
D	246	94
E	246	98

20. The table shows the radiation weighting factor w_R for a number of different radiations.

Type of radiation	Radiation weighting factor w_R
alpha particles	20
beta particles	1
neutrons	3
gamma rays	1
X-rays	0·1

Which of the following gives the greatest equivalent dose?

A $8\,\mu Gy$ of alpha particles

B $170\,\mu Gy$ of beta particles

C $56\,\mu Gy$ of neutrons

D $160\,\mu Gy$ of gamma rays

E $1500\,\mu Gy$ of X-rays

SECTION B begins on *Page eleven*

SECTION B

Write your answers to questions 21 to 29 in the answer book.

Marks

21. A van of mass 2600 kg moves down a slope which is inclined at 12° to the horizontal as shown.

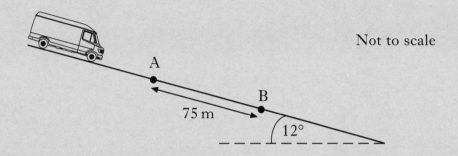

Not to scale

(a) Calculate the component of the van's weight parallel to the slope.   2

(b) A constant frictional force of 1400 N acts on the van as it moves down the slope.

Calculate the acceleration of the van.   2

(c) The speed of the van as it passes point **A** is $5.0 \, \text{m s}^{-1}$.
Point **B** is 75 m further down the slope.

Calculate the kinetic energy of the van at **B**.   3

(7)

[Turn over

Marks

22. A force sensor is used to investigate the impact of a ball as it bounces on a flat horizontal surface. The ball has a mass of 0·050 kg and is dropped vertically, from rest, through a height of 1·6 m as shown.

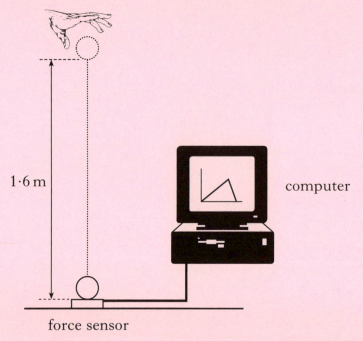

(a) The graph shows how the force on the ball varies with time during the impact.

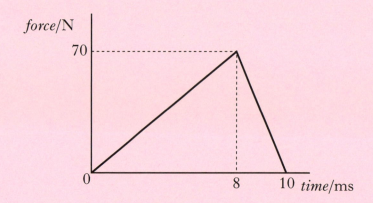

(i) Show by calculation that the magnitude of the impulse on the ball is 0·35 N s.   1

(ii) What is the magnitude and direction of the change in momentum of the ball?   1

(iii) The ball is travelling at 5·6 m s^{-1} just before it hits the force sensor. Calculate the speed of the ball just as it leaves the force sensor.   2

(b) Another ball of identical size and mass, but made of a harder material, is dropped from rest and from the same height onto the same force sensor. Sketch the force-time graph shown above and, on the same axes, sketch another graph to show how the force on the harder ball varies with time.

Numerical values are not required but you must label the graphs clearly.   2

(6)

Marks

23. A refrigerated cool box is being prepared to carry medical supplies in a hot country. The **internal** dimensions of the box are $0.30\,m \times 0.20\,m \times 0.50\,m$.

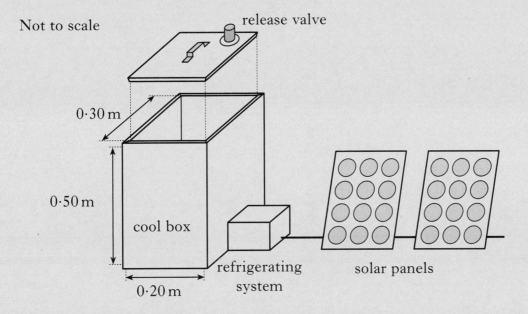

The lid is placed on the cool box with the release valve closed. An airtight seal is formed. When the lid is closed the air inside the cool box is at a temperature of $33\,°C$ and a pressure of $1.01 \times 10^5\,Pa$.

The refrigerating system then reduces the temperature of the air inside the cool box until it reaches its working temperature.

At this temperature the air inside is at a pressure of $9.05 \times 10^4\,Pa$.

(a) (i) Calculate the temperature of the air inside the cool box when it is at its working temperature. **2**

    (ii) Describe, using the kinetic model, how the decrease in temperature affects the air pressure inside the cool box. **2**

(b) (i) Atmospheric pressure is $1.01 \times 10^5\,Pa$.

       Show that the magnitude of the force on the lid due to the difference in air pressure between the inside and outside of the cool box is now $630\,N$. **2**

    (ii) The mass of the lid is $1.50\,kg$.

       Calculate the minimum force required to lift off the lid when the cool box is at its working temperature. **1**

    (iii) The release valve allows air to pass into or out of the cool box.

        Explain why this valve should be opened before lifting the lid. **1**

(c) The refrigerating system requires an average current of $0.80\,A$ at $12\,V$.

Each solar panel has a power output of $3.4\,W$ at $12\,V$.

Calculate the minimum number of solar panels needed to operate the refrigerating system. **2**

(10)

Marks

24. The diagram below shows the basic features of a proton accelerator. It is enclosed in an evacuated container.

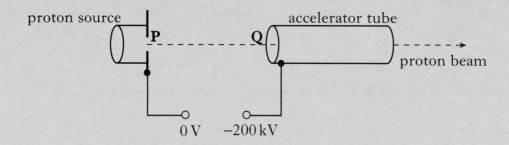

Protons released from the proton source start from rest at **P**.
A potential difference of 200 kV is maintained between **P** and **Q**.

(a) What is meant by the term *potential difference of 200 kV*?   1

(b) Explain why protons released at **P** are accelerated towards **Q**.   1

(c) Calculate:

   (i)   the work done on a proton as it accelerates from **P** to **Q**;   2

   (ii)   the speed of a proton as it reaches **Q**.   2

(d) The distance between **P** and **Q** is now halved.

What effect, if any, does this change have on the speed of a proton as it reaches **Q**? Justify your answer.   2

(8)

Marks

25. The 9·0 V battery in the circuit shown below has negligible internal resistance.

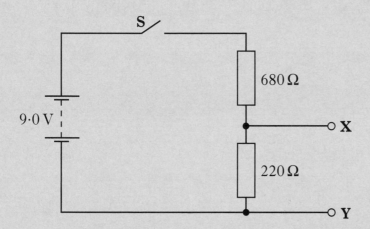

(a) Switch **S** is closed.
Calculate the potential difference between **X** and **Y**. 2

(b) Switch **S** is opened.

An uncharged 33 μF capacitor is connected between **X** and **Y** as shown.

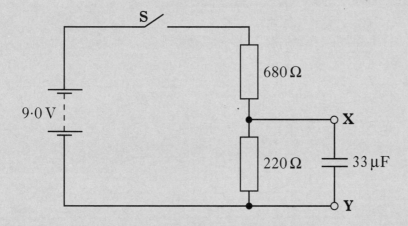

Switch **S** is then closed.

(i) Explain why work is done in charging the capacitor. 1

(ii) State the value of the maximum potential difference across the capacitor in this circuit. 1

(iii) Calculate the maximum energy stored in the capacitor. 2

(iv) Switch **S** is now opened.

Sketch a graph to show how the current through the 220 Ω resistor varies with time from the moment the switch is opened.

Numerical values are required only on the current axis. 2

(8)

[Turn over

Marks

26. A double beam oscilloscope has two inputs which allows two signals to be viewed on the screen at the same time.

A double beam oscilloscope is connected to the input terminals **P** and **Q** and the output terminals **R** and **S** of a box containing an operational amplifier circuit. The operational amplifier is operating in the inverting mode.

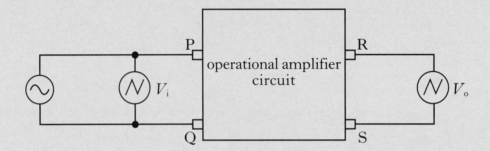

(a) The oscilloscope control settings and the two traces displayed on its screen are shown in the diagram.

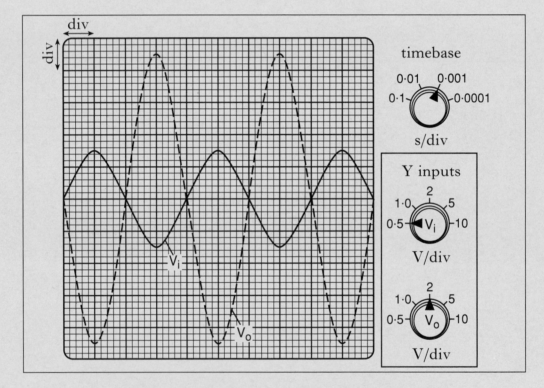

   (i) Calculate the frequency of the a.c. supply.   2

   (ii) Calculate the voltage gain of the amplifier circuit.   2

   (iii) Calculate the r.m.s. value of the output voltage of the amplifier circuit.   2

Marks

26. (continued)

(b) A student is given the task of altering the operational amplifier circuit inside the box to give a voltage gain of −4·7.

The following list shows resistor values available to the student.

Resistor value/kΩ
3·9
4·7
5·6
6·8
8·2
10
27
47
56

(i) Select suitable resistor values to produce a voltage gain of −4·7.   **1**

(ii) Copy the diagram shown below.

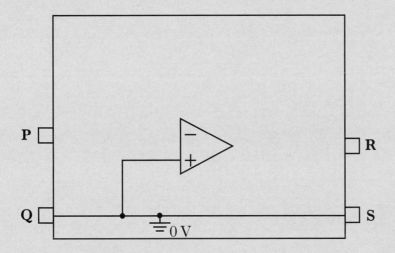

Complete the diagram, showing how your chosen resistors should be connected inside the box to complete the circuit.   **1**

(8)

[Turn over

Marks

27. (*a*) Light of frequency 6.7×10^{14} Hz is produced at the junction of a light emitting diode (LED).

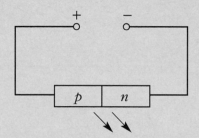

    (i) Describe how the movement of charges in a forward-biased LED produces light. Your description should include the terms: *electrons; holes; photons* and *junction*.   **1**

    (ii) (A) Calculate the wavelength of the light emitted from the LED.   **2**

        (B) Use information from the data sheet on *Page two* to deduce the colour of this light.   **1**

    (iii) The table below gives the values of the work function for three metals.

Metal	Work function/ J
caesium	3.4×10^{-19}
strontium	4.1×10^{-19}
magnesium	5.9×10^{-19}

Light from the LED is now incident on these metals in turn.

Show by calculation which of these metals, if any, release(s) photoelectrons with this light.   **3**

(*b*) Light from a different LED is passed through a grating as shown below.

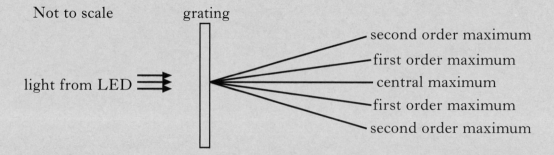

Light from this LED has a wavelength of 6.35×10^{-7} m. The spacing between lines in the grating is 5.0×10^{-6} m.

Calculate the angle between the central maximum and the **second** order maximum.   **2**

(9)

Marks

28. A student carries out an experiment to investigate how irradiance on a surface varies with distance from a small lamp.

Irradiance is measured with a light meter.

The distance between the small lamp and the light meter is measured with a metre stick.

The apparatus is set up as shown in a darkened laboratory.

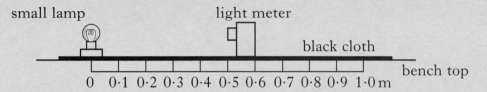

The following results are obtained.

Distance from source/ m	0·20	0·30	0·40	0·50
Irradiance/ units	675	302	170	108

(a) What is meant by the term *irradiance*?   1

(b) Use **all** the data to find the relationship between irradiance I and distance d from the source.   2

(c) What is the purpose of the black cloth on top of the bench?   1

(d) The small lamp is replaced by a laser.

Light from the laser is shone on to the light meter.

A reading is taken from the light meter when the distance between it and the laser is 0·50 m.

The distance is now increased to 1·00 m.

State how the new reading on the light meter compares with the one taken at 0·50 m.

Justify your answer.   2

(6)

[Turn over

Marks

29. (*a*) About one hundred years ago Rutherford designed an experiment to investigate the structure of the atom. He used a radioactive source to fire alpha particles at a thin gold foil target.

His two assistants, Geiger and Marsden, spent many hours taking readings from the detector as it was moved to different positions between **X** and **Y**.

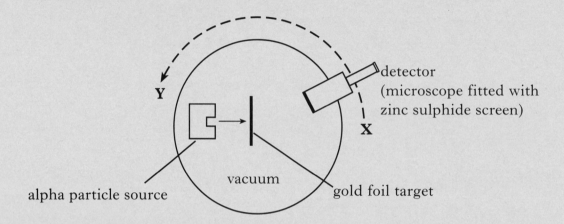

(i) How did the number of alpha particles detected at **X** compare with the number detected at **Y**?   **1**

(ii) State **one** conclusion Rutherford deduced from the results.   **1**

Marks

29.   (continued)

(*b*) A nuclear fission reaction is represented by the following statement.

$$^{235}_{92}\text{U} \;+\; ^{1}_{0}\text{n} \;\rightarrow\; ^{137}_{r}\text{Cs} \;+\; ^{s}_{37}\boldsymbol{T} \;+\; 4^{1}_{0}\text{n}$$

(i)   Is this a spontaneous or an induced reaction? You must justify your answer.   **1**

(ii) Determine the numbers represented by the letters *r* and *s* in the above reaction.   **1**

(iii) Use the data booklet to identify the element represented by *T*.   **1**

(iv) The masses of the nuclei and particles in the reaction are given below.

	Mass/kg
$^{235}_{92}\text{U}$	$390{\cdot}219 \times 10^{-27}$
$^{137}_{r}\text{Cs}$	$227{\cdot}292 \times 10^{-27}$
$^{s}_{37}\boldsymbol{T}$	$157{\cdot}562 \times 10^{-27}$
$^{1}_{0}\text{n}$	$1{\cdot}675 \times 10^{-27}$

Calculate the energy released in the reaction.   **3**

(8)

[END OF QUESTION PAPER]

[BLANK PAGE]

[BLANK PAGE]

X069/301

NATIONAL
QUALIFICATIONS
2007

WEDNESDAY, 16 MAY
1.00 PM – 3.30 PM

PHYSICS
HIGHER

Read Carefully

Reference may be made to the Physics Data Booklet.

1 All questions should be attempted.

Section A (questions 1 to 20)

2 Check that the answer sheet is for Physics Higher (Section A).

3 For this section of the examination you must use an **HB pencil** and, where necessary, an eraser.

4 Check that the answer sheet you have been given has **your name**, **date of birth**, **SCN** (Scottish Candidate Number) and **Centre Name** printed on it.

   Do not change any of these details.

5 If any of this information is wrong, tell the Invigilator immediately.

6 If this information is correct, **print** your name and seat number in the boxes provided.

7 There is **only one correct** answer to each question.

8 Any rough working should be done on the question paper or the rough working sheet, **not** on your answer sheet.

9 At the end of the exam, put the **answer sheet for Section A inside the front cover of your answer book**.

10 Instructions as to how to record your answers to questions 1–20 are given on page three.

Section B (questions 21 to 31)

11 Answer the questions numbered 21 to 31 in the answer book provided.

12 **All answers must be written clearly and legibly in ink**.

13 Fill in the details on the front of the answer book.

14 Enter the question number clearly in the margin of the answer book beside each of your answers to questions 21 to 31.

15 Care should be taken to give an appropriate number of significant figures in the final answers to calculations.

16 Where additional paper, eg square ruled paper, is used, write your name and SCN (Scottish Candidate Number) on it and place it inside the front cover of your answer booklet.

SCOTTISH
QUALIFICATIONS
AUTHORITY

DATA SHEET
COMMON PHYSICAL QUANTITIES

Quantity	Symbol	Value	Quantity	Symbol	Value
Speed of light in vacuum	c	$3 \cdot 00 \times 10^{8}$ m s^{-1}	Mass of electron	m_e	$9 \cdot 11 \times 10^{-31}$ kg
Magnitude of the charge on an electron	e	$1 \cdot 60 \times 10^{-19}$ C	Mass of neutron	m_n	$1 \cdot 675 \times 10^{-27}$ kg
Gravitational acceleration on Earth	g	$9 \cdot 8$ m s^{-2}	Mass of proton	m_p	$1 \cdot 673 \times 10^{-27}$ kg
Planck's constant	h	$6 \cdot 63 \times 10^{-34}$ J s			

REFRACTIVE INDICES
The refractive indices refer to sodium light of wavelength 589 nm and to substances at a temperature of 273 K.

Substance	Refractive index	Substance	Refractive index
Diamond	2·42	Water	1·33
Crown glass	1·50	Air	1·00

SPECTRAL LINES

Element	Wavelength/nm	Colour	Element	Wavelength/nm	Colour
Hydrogen	656	Red	Cadmium	644	Red
	486	Blue-green		509	Green
	434	Blue-violet		480	Blue
	410	Violet			
	397	Ultraviolet		*Lasers*	
	389	Ultraviolet	Element	Wavelength/nm	Colour
Sodium	589	Yellow	Carbon dioxide	9550 } 10590 }	Infrared
			Helium-neon	633	Red

PROPERTIES OF SELECTED MATERIALS

Substance	Density/ kg m^{-3}	Melting Point/ K	Boiling Point/ K
Aluminium	$2 \cdot 70 \times 10^{3}$	933	2623
Copper	$8 \cdot 96 \times 10^{3}$	1357	2853
Ice	$9 \cdot 20 \times 10^{2}$	273	. . . .
Sea Water	$1 \cdot 02 \times 10^{3}$	264	377
Water	$1 \cdot 00 \times 10^{3}$	273	373
Air	$1 \cdot 29$	. . . .	. . . .
Hydrogen	$9 \cdot 0 \ \times 10^{-2}$	14	20

The gas densities refer to a temperature of 273 K and a pressure of $1 \cdot 01 \times 10^{5}$ Pa.

SECTION A

For questions 1 to 20 in this section of the paper the answer to each question is either A, B, C, D or E. Decide what your answer is, then, using your pencil, put a horizontal line in the space provided—see the example below.

EXAMPLE

The energy unit measured by the electricity meter in your home is the

   A kilowatt-hour

   B ampere

   C watt

   D coulomb

   E volt.

The correct answer is **A**—kilowatt-hour. The answer **A** has been clearly marked in **pencil** with a horizontal line (see below).

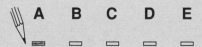

Changing an answer

If you decide to change your answer, carefully erase your first answer and, using your pencil, fill in the answer you want. The answer below has been changed to **E**.

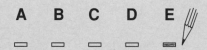

[Turn over

SECTION A

Answer questions 1–20 on the answer sheet.

1. Which row shows both quantities classified correctly?

	Scalar	Vector
A	weight	force
B	force	mass
C	mass	distance
D	distance	momentum
E	momentum	time

2. A ball is thrown vertically upwards and falls back to Earth. Neglecting air resistance, which velocity-time graph represents its motion?

A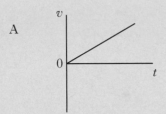

B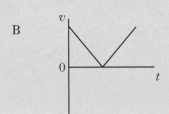

C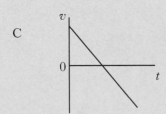

D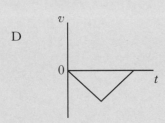

E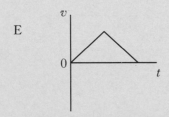

3. A person stands on a weighing machine in a lift. When the lift is at rest, the reading on the machine is 700 N. The lift now descends and its speed increases at a constant rate. The reading on the machine

 A is a constant value higher than 700 N

 B is a constant value lower than 700 N

 C continually increases from 700 N

 D continually decreases from 700 N

 E remains constant at 700 N.

4. Momentum can be measured in

 A $N\,kg^{-1}$

 B $N\,m$

 C $N\,m\,s^{-1}$

 D $kg\,m\,s^{-1}$

 E $kg\,m\,s^{-2}$.

5. A cannon of mass 2000 kg fires a cannonball of mass 5·00 kg.

 The cannonball leaves the cannon with a speed of $50{\cdot}0\,m\,s^{-1}$.

 The speed of the cannon immediately after firing is

 A $0{\cdot}125\,m\,s^{-1}$

 B $8{\cdot}00\,m\,s^{-1}$

 C $39{\cdot}9\,m\,s^{-1}$

 D $40{\cdot}1\,m\,s^{-1}$

 E $200\,m\,s^{-1}$.

6. The graph shows the force acting on an object of mass $5 \cdot 0$ kg.

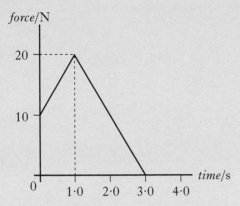

The change in the object's momentum is

A $\quad 7 \cdot 0$ kg m s^{-1}

B $\quad 30$ kg m s^{-1}

C $\quad 35$ kg m s^{-1}

D $\quad 60$ kg m s^{-1}

E $\quad 175$ kg m s^{-1}.

7. Which of the following gives the approximate relative spacings of molecules in ice, water and water vapour?

	Spacing of molecules in ice	Spacing of molecules in water	Spacing of molecules in water vapour
A	1	1	10
B	1	3	1
C	1	3	3
D	1	10	10
E	3	1	10

8. The element of an electric kettle has a resistance of 30 Ω. The kettle is connected to a mains supply. The r.m.s. voltage of this supply is 230 V. The peak value of the current in the kettle is

A $\quad 0 \cdot 13$ A

B $\quad 0 \cdot 18$ A

C $\quad 5 \cdot 4$ A

D $\quad 7 \cdot 7$ A

E $\quad 10 \cdot 8$ A.

9. Four resistors, each of resistance 20 Ω, are connected to a 60 V supply of negligible internal resistance, as shown.

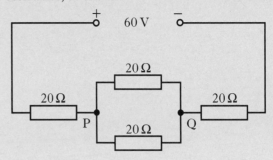

The potential difference across PQ is

A $\quad 12$ V

B $\quad 15$ V

C $\quad 20$ V

D $\quad 24$ V

E $\quad 30$ V.

10. A signal from a power supply is displayed on an oscilloscope.

The trace on the oscilloscope is shown.

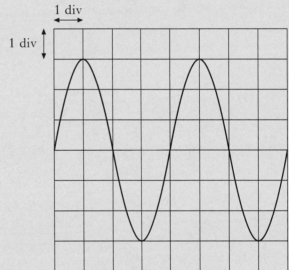

The time-base is set at $0 \cdot 01$ s/div and the Y-gain is set at $4 \cdot 0$ V/div.

Which row in the table shows the r.m.s. voltage and the frequency of the signal?

	r.m.s. voltage/V	frequency/Hz
A	$8 \cdot 5$	25
B	12	25
C	24	25
D	$8 \cdot 5$	50
E	12	50

11. A resistor and an ammeter are connected to a signal generator which has an output of constant amplitude and variable frequency.

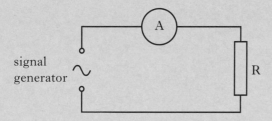

Which graph shows the relationship between the current I in the resistor and the output frequency f of the signal generator?

A
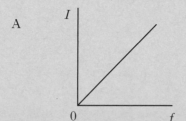

B
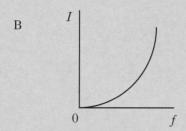

C
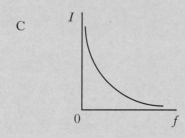

D
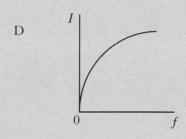

E
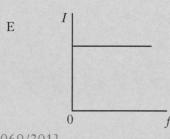

12. An oscilloscope is used to measure the frequency of the output voltage from an op-amp.

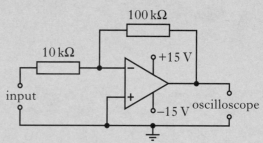

The input has a frequency of 280 Hz and a peak voltage of 0·5 V.

The frequency of the output voltage is

A   28 Hz

B   140 Hz

C   280 Hz

D   560 Hz

E   2800 Hz.

13. An op-amp circuit is set up as shown.

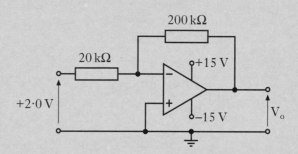

A voltage of +2·0 V is applied to the input. The voltage output, V_o, is approximately

A   +20 V

B   +15 V

C   −2·0 V

D   −15 V

E   −20 V.

14. The energy of a wave depends on its

A   amplitude

B   period

C   phase

D   speed

E   wavelength.

15. A ray of light travels from air into a glass prism. The refractive index of the glass is 1·50.

Which diagram shows the correct path of the ray?

A

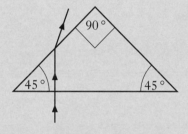

B

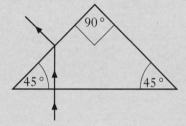

C

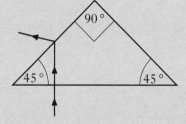

D

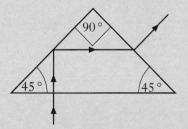

E

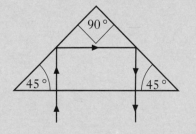

16. A beam of white light is passed through two optical components P and Q. Component P produces a number of spectra and component Q produces a spectrum as shown.

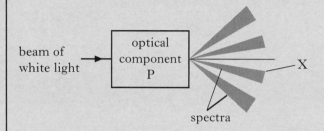

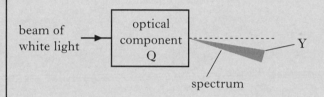

Which row in the table identifies the optical components and the colour of light seen at position X and position Y?

	Optical component P	Colour seen at X	Optical component Q	Colour seen at Y
A	grating	red	triangular prism	red
B	grating	red	triangular prism	violet
C	grating	violet	triangular prism	red
D	triangular prism	red	grating	violet
E	triangular prism	violet	grating	red

[Turn over

17. The diagram represents some electron transitions between energy levels in an atom.

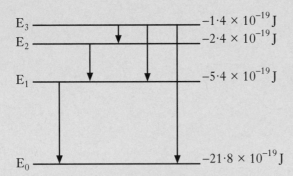

E_3 ——————— $-1\cdot4 \times 10^{-19}$ J
E_2 ——————— $-2\cdot4 \times 10^{-19}$ J

E_1 ——————— $-5\cdot4 \times 10^{-19}$ J

E_0 ——————— $-21\cdot8 \times 10^{-19}$ J

The radiation emitted with the shortest wavelength is produced by an electron making transition

A E_1 to E_0

B E_2 to E_1

C E_3 to E_2

D E_3 to E_1

E E_3 to E_0.

18. In the following circuit, component X is used to drive a motor.

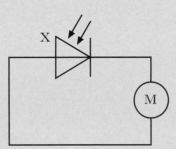

Which of the following gives the name of component X and its mode of operation?

	Name of component X	Mode of operation
A	light-emitting diode	photoconductive
B	light-emitting diode	photovoltaic
C	photodiode	photoconductive
D	photodiode	photovoltaic
E	op-amp	inverting

19. The classical experiment on the scattering of alpha particles from a thin gold foil suggested that

A   positive charges were evenly distributed throughout the atom

B   atomic nuclei were very small and positively charged

C   neutrons existed in the nucleus

D   alpha particles were helium nuclei

E   alpha particles were hydrogen nuclei.

20. A radioactive source produces a count rate of 2400 counts per second in a detector. When a lead plate of thickness 36 mm is placed between the source and the detector the count rate falls to 300 counts per second.

The half-value thickness of lead for this radiation is

A   4·5 mm

B   12 mm

C   36 mm

D   108 mm

E   288 mm.

[Turn over for SECTION B on *Page ten*

SECTION B

Write your answers to questions 21 to 31 in the answer book.

Marks

21. Competitors are racing remote control cars. The cars have to be driven over a precise route between checkpoints.

Checkpoint A

Each car is to travel from checkpoint A to checkpoint B by following these instructions.

"Drive 150 m due North, then drive 250 m on a bearing of 60° East of North (060)."

Car X takes 1 minute 6 seconds to follow these instructions exactly.

(a) By scale drawing or otherwise, find the displacement of checkpoint B from checkpoint A.

2

(b) Calculate the average velocity of car X from checkpoint A to checkpoint B.

2

(c) Car Y leaves A at the same time as car X.

Car Y follows exactly the same route at an average speed of $6\cdot5\,\mathrm{m\,s^{-1}}$.

Which car arrives first at checkpoint B?

Justify your answer with a calculation.

2

(d) State the displacement of checkpoint A from checkpoint B.

1

(7)

Marks

22. A fairground ride consists of rafts which slide down a slope into water.

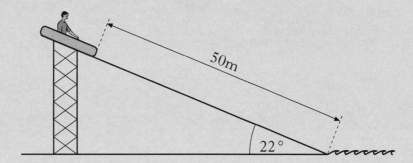

The slope is at an angle of 22° to the horizontal. Each raft has a mass of 8·0 kg. The length of the slope is 50 m.

A child of mass 52 kg sits in a raft at the top of the slope. The raft is released from rest. The child and raft slide together down the slope into the water. The force of friction between the raft and slope remains constant at 180 N.

(*a*) Calculate the component of weight, in newtons, of the child and raft down the slope.   **1**

(*b*) Show by calculation that the acceleration of the child and raft down the slope is 0·67 m s^{-2}.   **2**

(*c*) Calculate the speed of the child and raft at the bottom of the slope.   **2**

(*d*) A second child of smaller mass is released from rest in an identical raft at the same starting point. The force of friction is the same as before.

How does the speed of this child and raft at the bottom of the slope compare with the answer to part (*c*)?

Justify your answer.   **2**

(7)

[Turn over

Marks

23. A rigid cylinder contains $8 \cdot 0 \times 10^{-2}$ m³ of helium gas at a pressure of 750 kPa. Gas is released from the cylinder to fill party balloons.

rigid cylinder

party balloons

During the filling process, the temperature remains constant. When filled, each balloon holds 0·020 m³ of helium gas at a pressure of 125 kPa.

(a) Calculate the total volume of the helium gas when it is at a pressure of 125 kPa.   2

(b) Determine the maximum number of balloons which can be fully inflated by releasing gas from the cylinder.   2

(c) State how the density of the helium gas in an inflated balloon compares to the initial density of the helium gas inside the cylinder.

Justify your answer.   2

(6)

Marks

24. The apparatus shown in the diagram is designed to accelerate alpha particles.

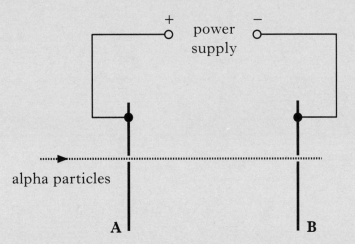

alpha particles

A   B

An alpha particle travelling at a speed of $2\cdot60 \times 10^6\,\text{m s}^{-1}$ passes through a hole in plate A. The mass of an alpha particle is $6\cdot64 \times 10^{-27}\,\text{kg}$ and its charge is $3\cdot2 \times 10^{-19}\,\text{C}$.

(a) When the alpha particle reaches plate B, its kinetic energy has increased to $3\cdot05 \times 10^{-14}\,\text{J}$.

Show that the work done on the alpha particle as it moves from plate A to plate B is $8\cdot1 \times 10^{-15}\,\text{J}$.   2

(b) Calculate the potential difference between plates A and B.   2

(c) The apparatus is now adapted to accelerate **electrons** from A to B through the same potential difference.

How does the increase in the kinetic energy of an electron compare with the increase in kinetic energy of the alpha particle in part (a)?

Justify your answer.   2

(6)

[Turn over

Marks

25. A power supply of e.m.f. E and internal resistance $2 \cdot 0 \, \Omega$ is connected as shown.

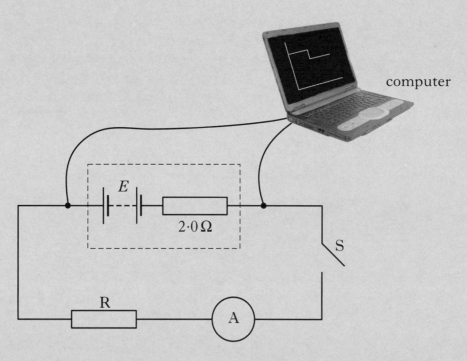

The computer connected to the apparatus displays a graph of potential difference against time.

The graph shows the potential difference across the terminals of the power supply for a short time before and after switch S is closed.

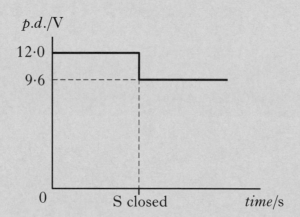

(a) State the e.m.f. of the power supply.   **1**

(b) Calculate:

    (i)   the reading on the ammeter after switch S is closed;   **2**

    (ii)   the resistance of resistor R.   **1**

Marks

25.   (continued)

(c)   Switch S is opened. A second identical resistor is now connected in parallel with R as shown.

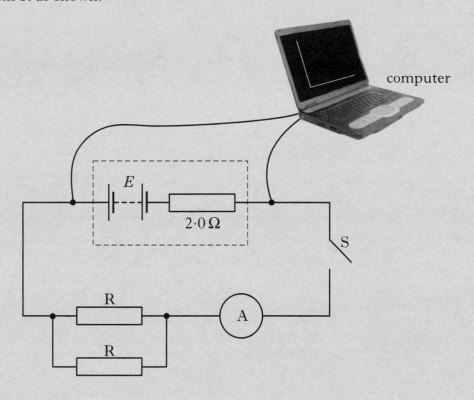

The computer is again connected in order to display a graph of potential difference against time.

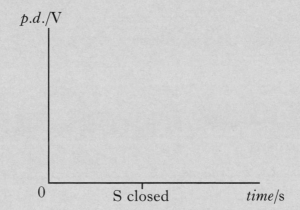

Copy and complete the new graph of potential difference against time showing the values of potential difference before and after switch S is closed.

2

(6)

[Turn over

Marks

26. An uncharged 2200 µF capacitor is connected in a circuit as shown.

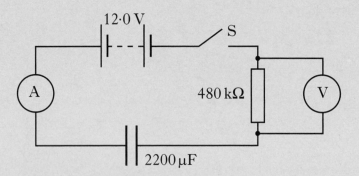

The battery has negligible internal resistance.

(a) Switch S is closed. Calculate the initial charging current.   2

(b) At one instant during the charging process the potential difference **across the resistor** is 3·8 V.

Calculate the charge stored in the capacitor at this instant.   3

(c) Calculate the **maximum** energy the capacitor stores in this circuit.   2

(7)

Marks

27. A Wheatstone bridge is used to measure the resistance of a thermistor as its temperature changes.

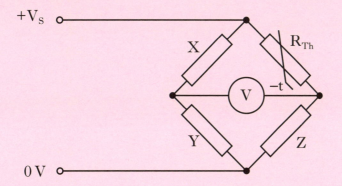

(a) The bridge is balanced when X = 2·2 kΩ, Y = 5·0 kΩ and Z = 750 Ω.

Calculate the resistance of the thermistor, R_{Th}, when the bridge is balanced.   **2**

(b) A student uses this bridge in a circuit to light an LED when the temperature in a greenhouse falls below a certain level.

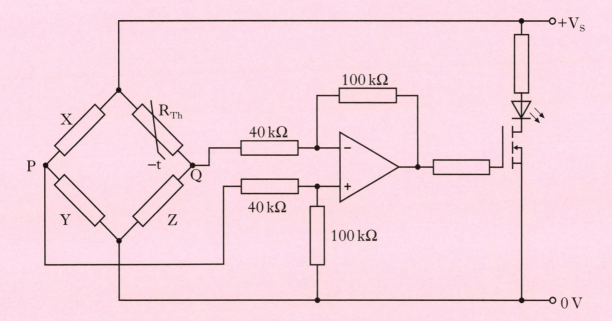

   (i) In which mode is the op-amp being used?   **1**

  (ii) As the temperature of the thermistor falls, its resistance increases.

Explain how this whole circuit operates to cause the LED to light when the temperature falls.   **2**

 (iii) At a certain temperature the output voltage of the op-amp is 3·0 V.

Calculate the potential difference between P and Q at this temperature.   **2**

(7)

[Turn over

Marks

28. An experiment to determine the wavelength of light from a laser is shown.

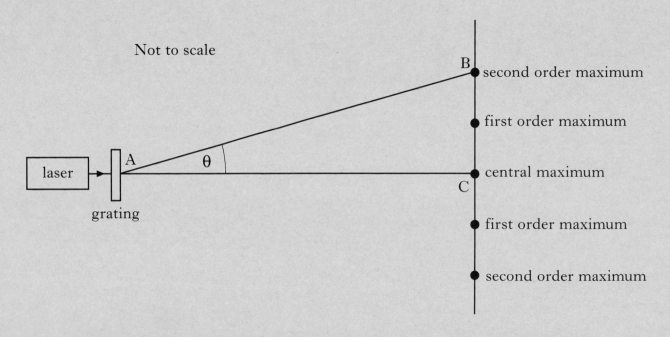

A **second** order maximum is observed at point B.

(a) Explain in terms of waves how a maximum is formed.   **1**

(b) Distance AB is measured six times.

The results are shown.

1·11 m   1·08 m   1·10 m   1·13 m   1·11 m   1·07 m

  (i) Calculate:

    (A) the mean value for distance AB;   **1**

    (B) the approximate random uncertainty in this value.   **1**

  (ii) Distance BC is measured as (270 ± 10) mm.

    Show whether AB or BC has the larger percentage uncertainty.   **2**

  (iii) The spacing between the lines on the grating is $4 \cdot 00 \times 10^{-6}$ m.

    Calculate the wavelength of the light from the laser.

    Express your answer in the form

        wavelength $\pm$ **absolute** uncertainty   **3**

                                                      (8)

Marks

29. A ray of red light is incident on a semicircular block of glass at the mid point of XY as shown.

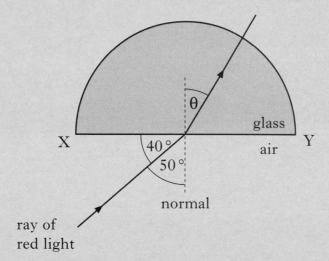

The refractive index of the block is 1·50 for this red light.

(a) Calculate angle θ shown on the diagram.   **2**

(b) The wavelength of the red light **in the glass** is 420 nm.

Calculate the wavelength of the light in air.   **2**

(c) The ray of red light is replaced by a ray of blue light incident at the same angle. The blue light enters the block at the same point.

Explain why the path taken by the blue light in the block is different to that taken by the red light.   **1**

(5)

[Turn over

Marks

30. A metal plate emits electrons when certain wavelengths of electromagnetic radiation are incident on it.

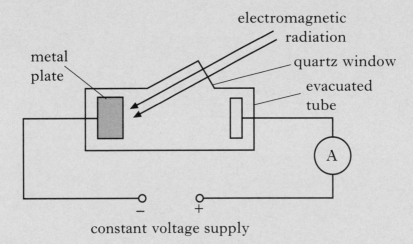

When light of wavelength 605 nm is incident on the metal plate, electrons are released with zero kinetic energy.

(a) Show that the work function of this metal is $3 \cdot 29 \times 10^{-19}$ J.   **2**

(b) The wavelength of the incident radiation is now altered. Photons of energy $5 \cdot 12 \times 10^{-19}$ J are incident on the metal plate.

  (i) Calculate the maximum kinetic energy of the electrons just as they leave the metal plate.   **1**

  (ii) The irradiance of this radiation on the metal plate is now decreased.

  State the effect this has on the ammeter reading.

  Justify your answer.   **2**

    (5)

Marks

31. (*a*) The following statement represents a nuclear reaction.

$$^{240}_{94}\text{Pu} \longrightarrow \, ^{236}_{92}\text{U} \; + \; ^{4}_{2}\text{He}$$

The table shows the masses of the particles involved in this reaction.

Particle	Mass/kg
$^{240}_{94}\text{Pu}$	$398{\cdot}626 \times 10^{-27}$
$^{236}_{92}\text{U}$	$391{\cdot}970 \times 10^{-27}$
$^{4}_{2}\text{He}$	$6{\cdot}645 \times 10^{-27}$

Calculate the energy released in this reaction.   **3**

(*b*) A technician is working with a radioactive source as shown.

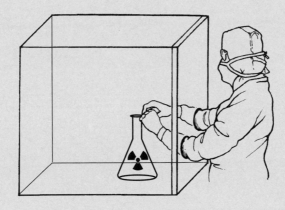

The technician's hands receive an absorbed dose at a rate of $4{\cdot}0\,\mu\text{Gy}\,\text{h}^{-1}$ for 2 hours. The radiation from the source has a radiation weighting factor of 3. Calculate the equivalent dose received by the technician's hands.   **3**

(6)

[*END OF QUESTION PAPER*]

[BLANK PAGE]

HIGHER

2008

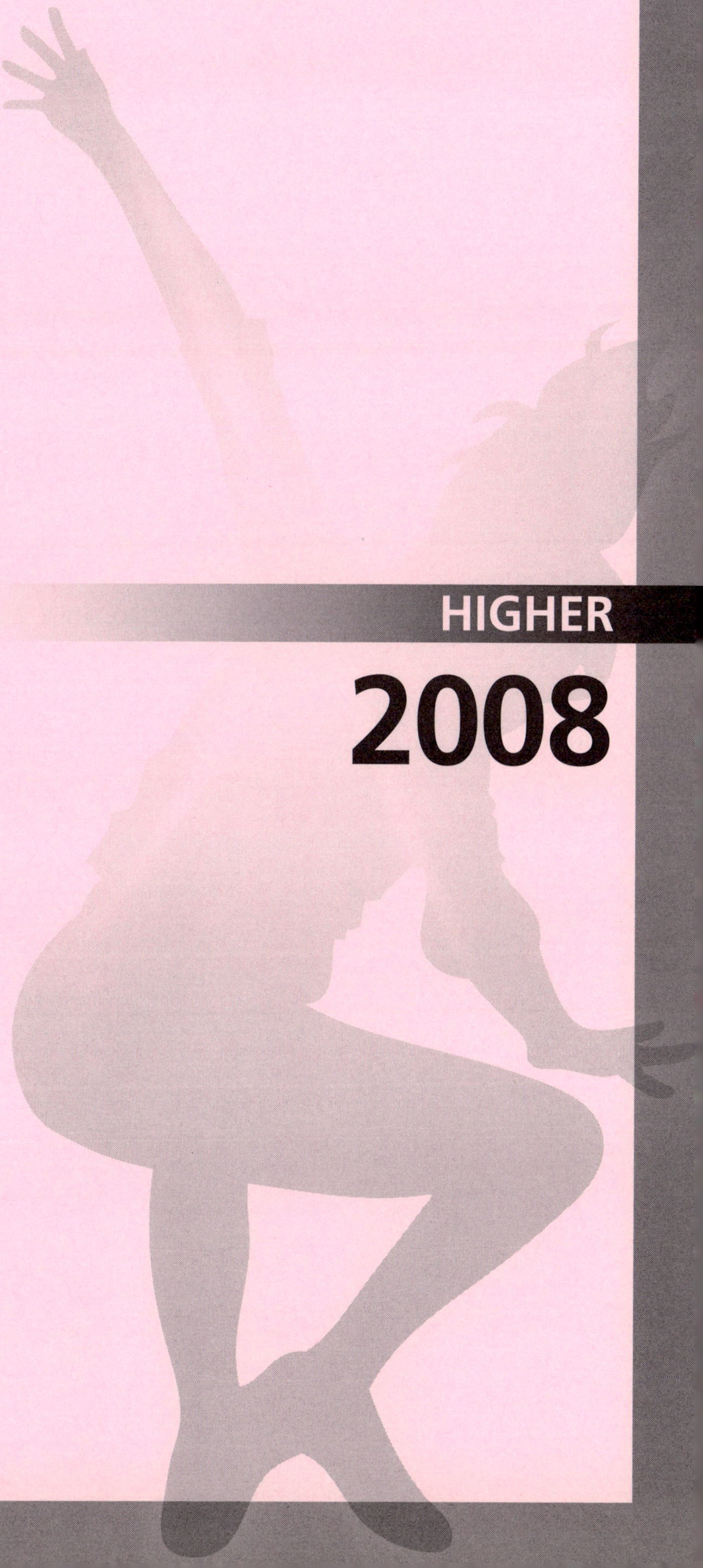

[BLANK PAGE]

X069/301

NATIONAL
QUALIFICATIONS
2008

FRIDAY, 23 MAY
1.00 PM – 3.30 PM

PHYSICS
HIGHER

Read Carefully

Reference may be made to the Physics Data Booklet.

1 All questions should be attempted.

Section A (questions 1 to 20)

2 Check that the answer sheet is for Physics Higher (Section A).

3 For this section of the examination you must use an **HB pencil** and, where necessary, an eraser.

4 Check that the answer sheet you have been given has **your name**, **date of birth**, **SCN** (Scottish Candidate Number) and **Centre Name** printed on it.

   Do not change any of these details.

5 If any of this information is wrong, tell the Invigilator immediately.

6 If this information is correct, **print** your name and seat number in the boxes provided.

7 There is **only one correct** answer to each question.

8 Any rough working should be done on the question paper or the rough working sheet, **not** on your answer sheet.

9 At the end of the exam, put the **answer sheet for Section A inside the front cover of your answer book**.

10 Instructions as to how to record your answers to questions 1–20 are given on page three.

Section B (questions 21 to 30)

11 Answer the questions numbered 21 to 30 in the answer book provided.

12 **All answers must be written clearly and legibly in ink**.

13 Fill in the details on the front of the answer book.

14 Enter the question number clearly in the margin of the answer book beside each of your answers to questions 21 to 30.

15 Care should be taken to give an appropriate number of significant figures in the final answers to calculations.

16 Where additional paper, eg square ruled paper, is used, write your name and SCN (Scottish Candidate Number) on it and place it inside the front cover of your answer booklet.

DATA SHEET
COMMON PHYSICAL QUANTITIES

Quantity	Symbol	Value	Quantity	Symbol	Value
Speed of light in vacuum	c	$3{\cdot}00 \times 10^{8}$ m s^{-1}	Mass of electron	m_e	$9{\cdot}11 \times 10^{-31}$ kg
Magnitude of the charge on an electron	e	$1{\cdot}60 \times 10^{-19}$ C	Mass of neutron	m_n	$1{\cdot}675 \times 10^{-27}$ kg
Gravitational acceleration on Earth	g	$9{\cdot}8$ m s^{-2}	Mass of proton	m_p	$1{\cdot}673 \times 10^{-27}$ kg
Planck's constant	h	$6{\cdot}63 \times 10^{-34}$ J s			

REFRACTIVE INDICES
The refractive indices refer to sodium light of wavelength 589 nm and to substances at a temperature of 273 K.

Substance	Refractive index	Substance	Refractive index
Diamond	2·42	Water	1·33
Crown glass	1·50	Air	1·00

SPECTRAL LINES

Element	Wavelength/nm	Colour	Element	Wavelength/nm	Colour
Hydrogen	656	Red	Cadmium	644	Red
	486	Blue-green		509	Green
	434	Blue-violet		480	Blue
	410	Violet			
	397	Ultraviolet		Lasers	
	389	Ultraviolet	Element	Wavelength/nm	Colour
			Carbon dioxide	9550 } 10590 }	Infrared
Sodium	589	Yellow	Helium-neon	633	Red

PROPERTIES OF SELECTED MATERIALS

Substance	Density/ kg m^{-3}	Melting Point/ K	Boiling Point/ K
Aluminium	$2{\cdot}70 \times 10^{3}$	933	2623
Copper	$8{\cdot}96 \times 10^{3}$	1357	2853
Ice	$9{\cdot}20 \times 10^{2}$	273	. . . .
Sea Water	$1{\cdot}02 \times 10^{3}$	264	377
Water	$1{\cdot}00 \times 10^{3}$	273	373
Air	$1{\cdot}29$	. . . .	. . . .
Hydrogen	$9{\cdot}0 \times 10^{-2}$	14	20

The gas densities refer to a temperature of 273 K and a pressure of $1{\cdot}01 \times 10^{5}$ Pa.

SECTION A

For questions 1 to 20 in this section of the paper the answer to each question is either A, B, C, D or E. Decide what your answer is, then, using your pencil, put a horizontal line in the space provided—see the example below.

EXAMPLE

The energy unit measured by the electricity meter in your home is the

    A kilowatt-hour

    B ampere

    C watt

    D coulomb

    E volt.

The correct answer is **A**—kilowatt-hour. The answer **A** has been clearly marked in **pencil** with a horizontal line (see below).

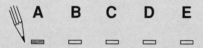

Changing an answer

If you decide to change your answer, carefully erase your first answer and, using your pencil, fill in the answer you want. The answer below has been changed to **E**.

[Turn over

SECTION A

Answer questions 1–20 on the answer sheet.

1. Which row in the table is correct?

	Scalar	*Vector*
A	distance	work
B	weight	acceleration
C	velocity	displacement
D	mass	momentum
E	speed	time

2. A javelin is thrown at 60° to the horizontal with a speed of 20 m s^{-1}.

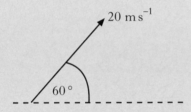

The javelin is in flight for 3·5 s.
Air resistance is negligible.
The horizontal distance the javelin travels is

A   35·0 m

B   60·6 m

C   70·0 m

D   121 m

E   140 m.

3. Two boxes on a frictionless horizontal surface are joined together by a string. A constant horizontal force of 12 N is applied as shown.

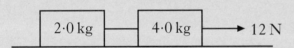

The tension in the string joining the two boxes is

A   2·0 N

B   4·0 N

C   6·0 N

D   8·0 N

E   12 N.

4. The total mass of a motorcycle and rider is 250 kg. During braking, they are brought to rest from a speed of 16·0 m s^{-1} in a time of 10·0 s.

The maximum energy which could be converted to heat in the brakes is

A   2000 J

B   4000 J

C   32 000 J

D   40 000 J

E   64 000 J.

5. A shell of mass 5·0 kg is travelling horizontally with a speed of 200 m s^{-1}. It explodes into two parts. One part of mass 3·0 kg continues in the original direction with a speed of 100 m s^{-1}.

The other part also continues in this same direction. Its speed is

A   150 m s^{-1}

B   200 m s^{-1}

C   300 m s^{-1}

D   350 m s^{-1}

E   700 m s^{-1}.

6. The graph shows the force which acts on an object over a time interval of 8 seconds.

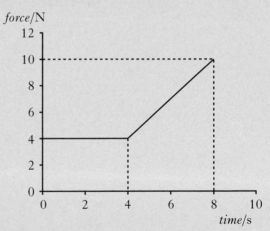

The momentum gained by the object during this 8 seconds is

A   12 kg m s^{-1}

B   32 kg m s^{-1}

C   44 kg m s^{-1}

D   52 kg m s^{-1}

E   72 kg m s^{-1}.

7. One pascal is equivalent to

A   1 N m

B   1 N m^2

C   1 N m^3

D   1 N m^{-2}

E   1 N m^{-3}.

8. An electron is accelerated from rest through a potential difference of 2·0 kV.

The kinetic energy gained by the electron is

A $8·0 \times 10^{-23}$ J

B $8·0 \times 10^{-20}$ J

C $3·2 \times 10^{-19}$ J

D $1·6 \times 10^{-16}$ J

E $3·2 \times 10^{-16}$ J.

9. The e.m.f. of a battery is

A   the total energy supplied by the battery

B   the voltage lost due to the internal resistance of the battery

C   the total charge which passes through the battery

D   the number of coulombs of charge passing through the battery per second

E   the energy supplied to each coulomb of charge passing through the battery.

10. The diagram shows the trace on an oscilloscope when an alternating voltage is applied to its input.

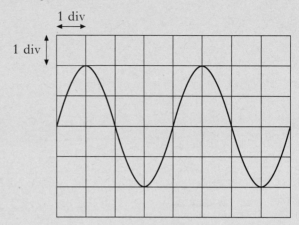

The timebase is set at 5 ms/div and the Y-gain is set at 10 V/div.

Which row in the table gives the peak voltage and the frequency of the signal?

	Peak voltage/V	Frequency/Hz
A	7·1	20
B	14	50
C	20	20
D	20	50
E	40	50

[Turn over

11. A resistor is connected to an a.c. supply as shown.

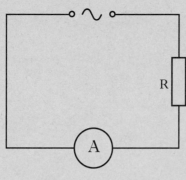

a.c. ammeter

The supply has a constant peak voltage, but its frequency can be varied.

The frequency is steadily increased from 50 Hz to 5000 Hz.

The reading on the a.c. ammeter

A   remains constant

B   decreases steadily

C   increases steadily

D   increases then decreases

E   decreases then increases.

12. An ideal op-amp is connected as shown.

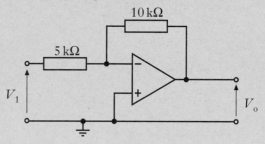

The graph shows how the input voltage, V_1, varies with time.

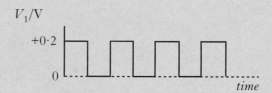

Which graph shows how the output voltage, V_o, varies with time?

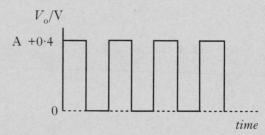

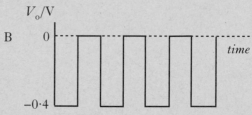

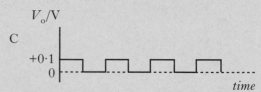

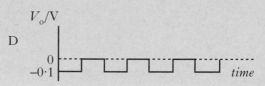

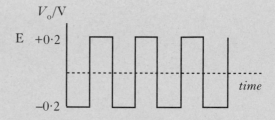

13. Which of the following proves that light is transmitted as waves?

A   Light has a high velocity.

B   Light can be reflected.

C   Light irradiance reduces with distance.

D   Light can be refracted.

E   Light can produce interference patterns.

14. A source of microwaves of wavelength λ is placed behind two slits, R and S.

A microwave detector records a maximum when it is placed at P.

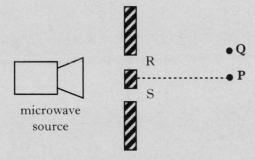

microwave source

The detector is moved and the **next** maximum is recorded at Q.

The path difference (SQ – RQ) is

A   0

B $\dfrac{\lambda}{2}$

C λ

D $\dfrac{3\lambda}{2}$

E 2λ.

15. A student makes five separate measurements of the diameter of a lens.

These measurements are shown in the table.

Diameter of lens/mm	22·5	22·6	22·4	22·6	22·9

The approximate random uncertainty in the mean value of the diameter is

A   0·1 mm

B   0·2 mm

C   0·3 mm

D   0·4 mm

E   0·5 mm.

16. The value of the absolute refractive index of diamond is 2·42.

The critical angle for diamond is

A   0·413°

B   24·4°

C   42·0°

D   65·6°

E   90·0°.

[Turn over

17. Part of the energy level diagram for an atom is shown.

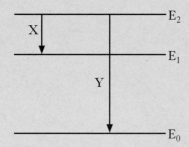

X and Y represent two possible electron transitions.
Which of the following statements is/are correct?

I Transition Y produces photons of higher frequency than transition X.

II Transition X produces photons of longer wavelength than transition Y.

III When an electron is in the energy level E_0, the atom is ionised.

A   I only

B   I and II only

C   I and III only

D   II and III only

E   I, II and III

18. The letters **X**, **Y** and **Z** represent three missing words from the following passage.

Materials can be divided into three broad categories according to their electrical resistance.

.................**X**............. *have a very high resistance.*

.................**Y**............. *have a high resistance in their pure form but when small amounts of certain impurities are added, the resistance decreases.*

.................**Z**............. *have a low resistance.*

Which row in the table shows the missing words?

	X	Y	Z
A	conductors	insulators	semi-conductors
B	semi-conductors	insulators	conductors
C	insulators	semi-conductors	conductors
D	conductors	semi-conductors	insulators
E	insulators	conductors	semi-conductors

19. Compared with a proton, an alpha particle has

A   twice the mass and twice the charge

B   twice the mass and the same charge

C   four times the mass and twice the charge

D   four times the mass and the same charge

E   twice the mass and four times the charge.

20. For the nuclear decay shown, which row of the table gives the correct values of x, y and z?

$$^{214}_{x}\text{Pb} \longrightarrow {}^{y}_{83}\text{Bi} + {}^{0}_{z}\text{e}$$

	x	y	z
A	85	214	2
B	84	214	1
C	83	210	4
D	82	214	−1
E	82	210	−1

[Turn over for SECTION B on *Page ten*

SECTION B

Write your answers to questions 21 to 30 in the answer book.

Marks

21. To test the braking system of cars, a test track is set up as shown.

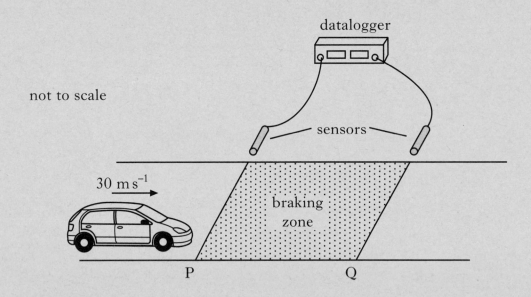

The sensors are connected to a datalogger which records the speed of a car at both P and Q.

A car is driven at a constant speed of 30 m s^{-1} until it reaches the start of the braking zone at P. The brakes are then applied.

(a) In one test, the datalogger records the speed at P as 30 m s^{-1} and the speed at Q as 12 m s^{-1}. The car slows down at a constant rate of $9 \cdot 0 \text{ m s}^{-2}$ between P and Q.

Calculate the length of the braking zone.   **2**

(b) The test is repeated. The same car is used but now with passengers in the car. The speed at P is again recorded as 30 m s^{-1}.

The same braking force is applied to the car as in part (a).

How does the speed of the car at Q compare with its speed at Q in part (a)? Justify your answer.   **2**

Marks

21.   (continued)

(c)   The brake lights of the car consist of a number of very bright LEDs.

An LED from the brake lights is forward biased by connecting it to a 12 V
car battery as shown.

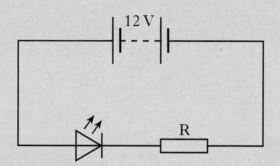

The battery has negligible internal resistance.

   (i)   Explain, in terms of charge carriers, how the LED emits light.   **1**

   (ii)   The LED is operating at its rated values of 5·0 V and 2·2 W.

Calculate the value of resistor R.   **3**

(8)

[Turn over

Marks

22. A crate of mass 40·0 kg is pulled up a slope using a rope.

The slope is at an angle of 30° to the horizontal.

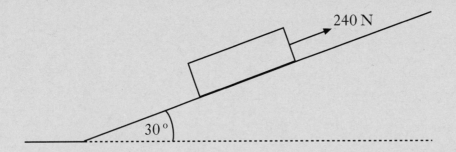

A force of 240 N is applied to the crate parallel to the slope.

The crate moves at a constant speed of 3·0 m s^{-1}.

(*a*) (i) Calculate the component of the weight of the crate acting parallel to the slope. **2**

   (ii) Calculate the frictional force acting on the crate. **2**

(*b*) As the crate is moving up the slope, the rope snaps.

The graph shows how the velocity of the crate changes from the moment the rope snaps.

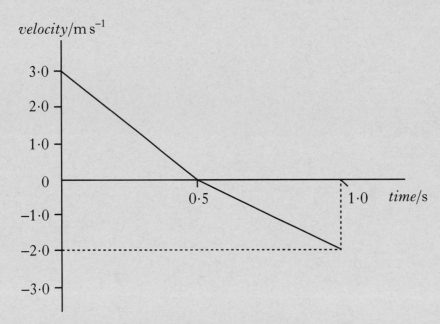

   (i) Describe the motion of the crate during the first 0·5 s after the rope snaps. **1**

Marks

22. **(*b*) (continued)**

(ii) Copy the axes shown below and sketch the graph to show the acceleration of the crate between 0 and 1·0 s.

Appropriate numerical values are also required on the acceleration axis.

2

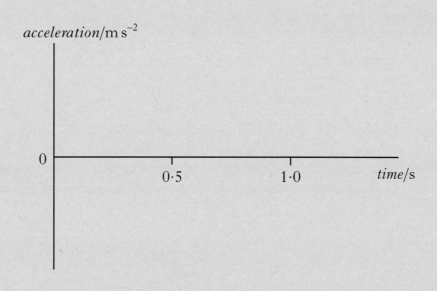

(iii) Explain, in terms of the forces acting on the crate, why the magnitude of the acceleration changes at 0·5 s.

2

(9)

[Turn over

Marks

23. A cylinder of compressed oxygen gas is in a laboratory.

(a) The oxygen inside the cylinder is at a pressure of $2 \cdot 82 \times 10^6$ Pa and a temperature of $19 \cdot 0\,°C$.

The cylinder is now moved to a storage room where the temperature is $5 \cdot 0\,°C$.

(i) Calculate the pressure of the oxygen inside the cylinder when its temperature is $5 \cdot 0\,°C$.   **2**

(ii) What effect, if any, does this decrease in temperature have on the density of the oxygen in the cylinder?

Justify your answer.   **2**

(b) (i) The volume of oxygen inside the cylinder is $0 \cdot 030$ m^3.

The density of the oxygen inside the cylinder is $37 \cdot 6$ kg m^{-3}.

Calculate the mass of oxygen in the cylinder.   **2**

(ii) The valve on the cylinder is opened slightly so that oxygen is gradually released.

The temperature of the oxygen inside the cylinder remains constant.

Explain, in terms of particles, why the pressure of the gas inside the cylinder decreases.   **1**

(iii) After a period of time, the pressure of the oxygen inside the cylinder reaches a constant value of $1 \cdot 01 \times 10^5$ Pa. The valve remains open.

Explain why the pressure does not decrease below this value.   **1**

(8)

Marks

24. Electrically heated gloves are used by skiers and climbers to provide extra warmth.

(a) Each glove has a heating element of resistance $3 \cdot 6 \, \Omega$.

Two cells, each of e.m.f. $1 \cdot 5 \, V$ and internal resistance $0 \cdot 20 \, \Omega$, are used to operate the heating element.

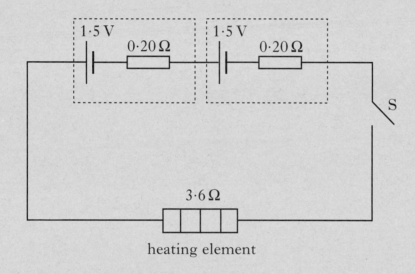

heating element

Switch S is closed.

(i) Determine the value of the total circuit resistance.   1

(ii) Calculate the current in the heating element.   2

(iii) Calculate the power output of the heating element.   2

(b) When in use, the internal resistance of each cell gradually increases.

What effect, if any, does this have on the power output of the heating element?

Justify your answer.   2

    (7)

[Turn over

Marks

25. (*a*) State what is meant by the term *capacitance*.   **1**

(*b*) An uncharged capacitor, C, is connected in a circuit as shown.

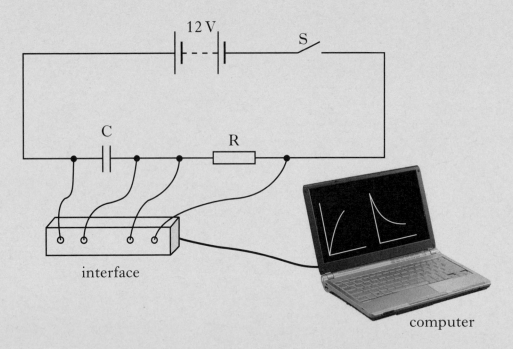

The 12 V battery has negligible internal resistance.

Switch S is closed and the capacitor begins to charge.

The interface measures the current in the circuit and the potential difference (p.d.) across the capacitor. These measurements are displayed as graphs on the computer.

Graph 1 shows the p.d. across the capacitor for the first 0·40 s of charging.

Graph 2 shows the current in the circuit for the first 0·40 s of charging.

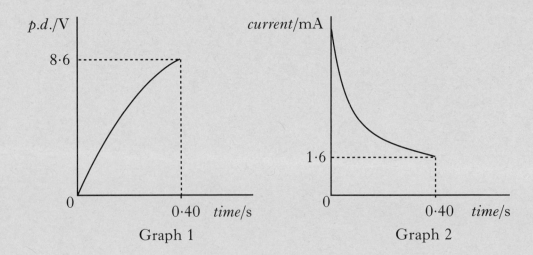

Marks

25. (b) (continued)

   (i)   Determine the p.d. **across resistor R** at 0·40 s.   1

   (ii) Calculate the resistance of R.   2

   (iii) The capacitor takes 2·2 seconds to charge fully.

         At that time it stores 10·8 mJ of energy.

         Calculate the capacitance of the capacitor.   3

(c) The capacitor is now discharged.
    A second, identical resistor is connected in the circuit as shown.

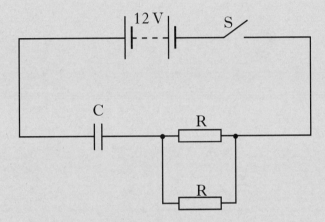

Switch S is closed.

Is the time taken for the capacitor to fully charge less than, equal to, or greater than the time taken to fully charge in part (b)?

Justify your answer.   2

(9)

[Turn over

Marks

26. The graph shows how the resistance of an LDR changes with the irradiance of light incident on it.

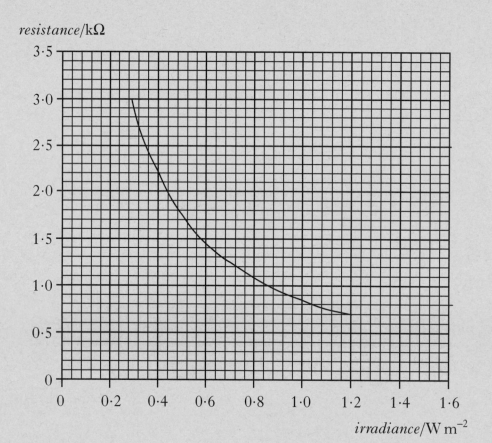

(*a*) The LDR is connected in the following bridge circuit.

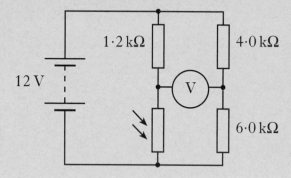

Determine the value of irradiance at which the bridge is balanced.

Show clearly how you arrive at your answer.   3

Marks

26.   (continued)

(*b*) The LDR is now mounted on the outside of a car to monitor light level. It forms part of a circuit which provides an indication for the driver to switch on the headlamps.
The circuit is shown below.

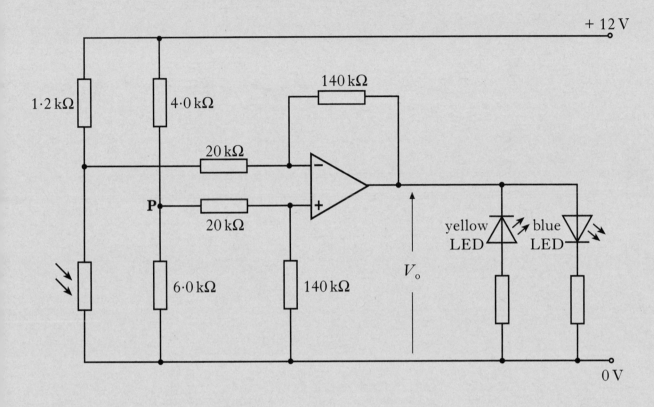

The LEDs inside the car indicate whether the headlamps should be on or off.

(i)   At a particular value of irradiance the resistance of the LDR is $2 \cdot 0\,\text{k}\Omega$.

Show that the potential difference across the LDR in the circuit is $7 \cdot 5\,\text{V}$.   1

(ii)   The potential at point P in the circuit is $7 \cdot 2\,\text{V}$.

Calculate the output voltage, V_{o}, of the op-amp at this light level.   2

(iii)   Which LED(s) is/are lit at this value of output voltage?

Justify your answer.   2

(8)

[Turn over

Marks

27. (*a*) A ray of red light of frequency 4.80×10^{14} Hz is incident on a glass lens as shown.

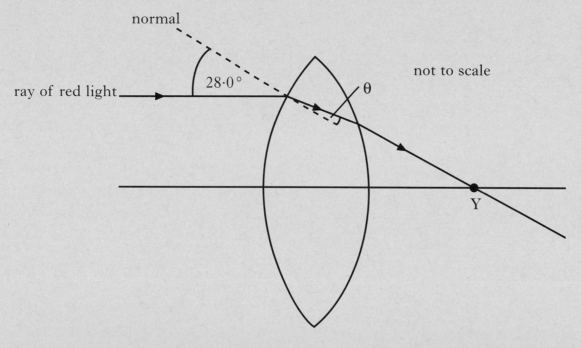

The ray passes through point Y after leaving the lens.

The refractive index of the glass is 1·61 for this red light.

  (i) Calculate the value of the angle θ shown in the diagram.   **2**

  (ii) Calculate the wavelength of this light inside the lens.   **3**

(*b*) The ray of red light is now replaced by a ray of blue light.

The ray is incident on the lens at the same point as in part (*a*).

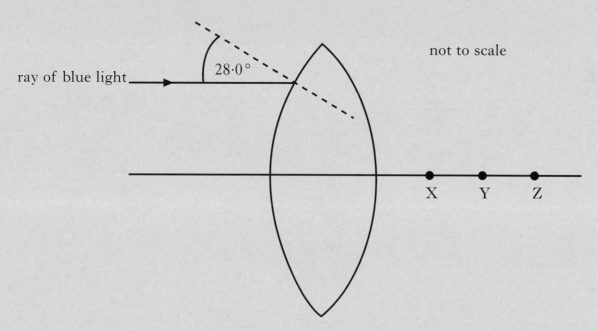

Through which point, X, Y or Z, will this ray pass after leaving the lens?

You must justify your answer.   **1**

(6)

Marks

28. The diagram shows a light sensor connected to a voltmeter.

A small lamp is placed in front of the sensor.

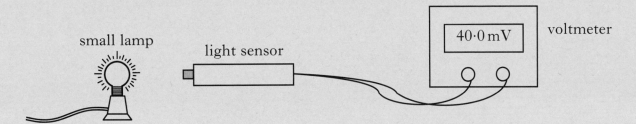

The reading on the voltmeter is 20 mV for each 1·0 mW of power incident on the sensor.

(a) The reading on the voltmeter is 40·0 mV.

The area of the light sensor is $8·0 \times 10^{-5} \, m^2$.

Calculate the irradiance of light on the sensor.   3

(b) The small lamp is replaced by a different source of light.

Using this new source, a student investigates how irradiance varies with distance.

The results are shown.

Distance/m	0·5	0·7	0·9
Irradiance/W m^{-2}	1·1	0·8	0·6

Can this new source be considered to be a point source of light?

Use **all** the data to justify your answer.   2

                                                                      (5)

[Turn over

Marks

29. To explain the photoelectric effect, light can be considered as consisting of tiny bundles of energy. These bundles of energy are called photons.

(a) Sketch a graph to show the relationship between photon energy and frequency.   **1**

(b) Photons of frequency $6 \cdot 1 \times 10^{14}$ Hz are incident on the surface of a metal.

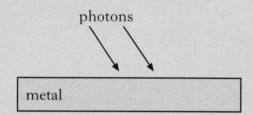

This releases photoelectrons from the surface of the metal.

The maximum kinetic energy of any of these photoelectrons is $6 \cdot 0 \times 10^{-20}$ J.

Calculate the work function of the metal.   **3**

(c) The irradiance due to these photons on the surface of the metal is now reduced.

Explain why the maximum kinetic energy of each photoelectron is unchanged.   **1**

(5)

Marks

30. (*a*) A technician is carrying out an experiment on the absorption of gamma radiation.

The radioactive source used has a long half-life and emits only gamma radiation. The activity of the source is 12 kBq.

   (i) State what is meant by an *activity of 12 kBq*.   **1**

   (ii) The table shows the half-value thicknesses of aluminium and lead for gamma radiation.

Material	Half-value thickness/mm
aluminium (Al)	60
lead (Pb)	15

The technician sets up the following apparatus.

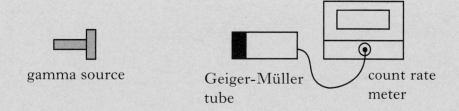

gamma source   Geiger-Müller tube   count rate meter

The count rate, when corrected for background radiation, is 800 counts per second.

Samples of aluminium and lead are now placed between the source and detector as shown.

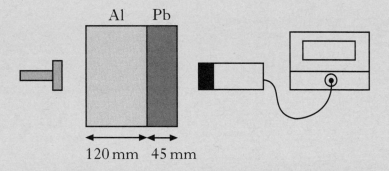

Al   Pb

120 mm   45 mm

Determine the new corrected count rate.   **2**

[Turn over for Question 30 (*b*) on *Page twenty-four*

Marks

30.   (continued)

(*b*)   X-ray scanners are used as part of airport security. A beam of X-rays scans the luggage as it passes through the scanner.

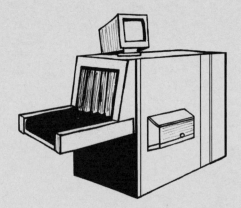

A baggage handler sometimes puts a hand inside the scanner to clear blockages.

The hand receives an average absorbed dose of $0.030\,\mu$Gy each time this occurs.

The radiation weighting factor for X-rays is 1.

(i)   State the average equivalent dose received by the hand on each occasion.   **1**

(ii)   The occupational exposure limit for a hand is $60\,\mu$Sv per hour.

Calculate how many times the baggage handler would have to put a hand into the scanner in one hour to reach this limit.   **1**

(5)

[END OF QUESTION PAPER]

2009

[BLANK PAGE]

X069/301

NATIONAL QUALIFICATIONS 2009	TUESDAY, 26 MAY 1.00 PM – 3.30 PM	PHYSICS HIGHER

Read Carefully

Reference may be made to the Physics Data Booklet.

1 All questions should be attempted.

Section A (questions 1 to 20)

2 Check that the answer sheet is for Physics Higher (Section A).

3 For this section of the examination you must use an **HB pencil** and, where necessary, an eraser.

4 Check that the answer sheet you have been given has **your name**, **date of birth**, **SCN** (Scottish Candidate Number) and **Centre Name** printed on it.

Do not change any of these details.

5 If any of this information is wrong, tell the Invigilator immediately.

6 If this information is correct, **print** your name and seat number in the boxes provided.

7 There is **only one correct** answer to each question.

8 Any rough working should be done on the question paper or the rough working sheet, **not** on your answer sheet.

9 At the end of the exam, put the **answer sheet for Section A inside the front cover of your answer book**.

10 Instructions as to how to record your answers to questions 1–20 are given on page three.

Section B (questions 21 to 30)

11 Answer the questions numbered 21 to 30 in the answer book provided.

12 **All answers must be written clearly and legibly in ink**.

13 Fill in the details on the front of the answer book.

14 Enter the question number clearly in the margin of the answer book beside each of your answers to questions 21 to 30.

15 Care should be taken to give an appropriate number of significant figures in the final answers to calculations.

16 Where additional paper, eg square ruled paper, is used, write your name and SCN (Scottish Candidate Number) on it and place it inside the front cover of your answer booklet.

DATA SHEET
COMMON PHYSICAL QUANTITIES

Quantity	Symbol	Value	Quantity	Symbol	Value
Speed of light in vacuum	c	$3{\cdot}00 \times 10^8$ m s^{-1}	Mass of electron	m_e	$9{\cdot}11 \times 10^{-31}$ kg
Magnitude of the charge on an electron	e	$1{\cdot}60 \times 10^{-19}$ C	Mass of neutron	m_n	$1{\cdot}675 \times 10^{-27}$ kg
Gravitational acceleration on Earth	g	$9{\cdot}8$ m s^{-2}	Mass of proton	m_p	$1{\cdot}673 \times 10^{-27}$ kg
Planck's constant	h	$6{\cdot}63 \times 10^{-34}$ J s			

REFRACTIVE INDICES
The refractive indices refer to sodium light of wavelength 589 nm and to substances at a temperature of 273 K.

Substance	Refractive index	Substance	Refractive index
Diamond	2·42	Water	1·33
Crown glass	1·50	Air	1·00

SPECTRAL LINES

Element	Wavelength/nm	Colour	Element	Wavelength/nm	Colour
Hydrogen	656	Red	Cadmium	644	Red
	486	Blue-green		509	Green
	434	Blue-violet		480	Blue
	410	Violet		Lasers	
	397	Ultraviolet	Element	Wavelength/nm	Colour
	389	Ultraviolet	Carbon dioxide	9550 / 10590	Infrared
Sodium	589	Yellow	Helium-neon	633	Red

PROPERTIES OF SELECTED MATERIALS

Substance	Density/ kg m^{-3}	Melting Point/ K	Boiling Point/ K
Aluminium	$2{\cdot}70 \times 10^3$	933	2623
Copper	$8{\cdot}96 \times 10^3$	1357	2853
Ice	$9{\cdot}20 \times 10^2$	273	....
Sea Water	$1{\cdot}02 \times 10^3$	264	377
Water	$1{\cdot}00 \times 10^3$	273	373
Air	1·29	....	....
Hydrogen	$9{\cdot}0 \times 10^{-2}$	14	20

The gas densities refer to a temperature of 273 K and a pressure of $1{\cdot}01 \times 10^5$ Pa.

SECTION A

For questions 1 to 20 in this section of the paper the answer to each question is either A, B, C, D or E. Decide what your answer is, then, using your pencil, put a horizontal line in the space provided—see the example below.

EXAMPLE

The energy unit measured by the electricity meter in your home is the

    A kilowatt-hour

    B ampere

    C watt

    D coulomb

    E volt.

The correct answer is **A**—kilowatt-hour. The answer **A** has been clearly marked in **pencil** with a horizontal line (see below).

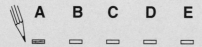

Changing an answer

If you decide to change your answer, carefully erase your first answer and, using your pencil, fill in the answer you want. The answer below has been changed to **E**.

[Turn over

SECTION A

Answer questions 1–20 on the answer sheet.

1. Which of the following contains one vector and one scalar quantity?

   A   power; speed

   B   force; kinetic energy

   C   momentum; velocity

   D   work; potential energy

   E   displacement; acceleration

2. The following velocity-time graph represents the vertical motion of a ball.

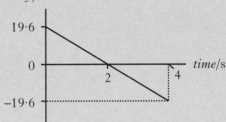

   Which of the following acceleration-time graphs represents the same motion?

   A
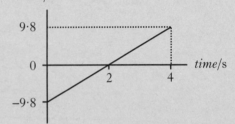

   B
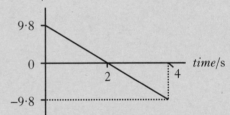

   C
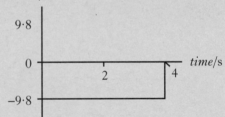

   D
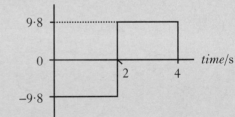

   E
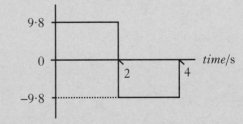

3. A box of weight 120 N is placed on a smooth horizontal surface.

A force of 20 N is applied to the box as shown.

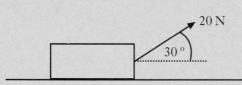

The box is pulled a distance of 50 m along the surface.

The work done in pulling the box is

A   500 J

B   866 J

C   1000 J

D   6000 J

E   6866 J.

4. A skydiver of total mass 85 kg is falling vertically.

At one point during the fall, the air resistance on the skydiver is 135 N.

The acceleration of the skydiver at this point is

A $0 \cdot 6 \, \mathrm{m \, s^{-2}}$

B $1 \cdot 6 \, \mathrm{m \, s^{-2}}$

C $6 \cdot 2 \, \mathrm{m \, s^{-2}}$

D $8 \cdot 2 \, \mathrm{m \, s^{-2}}$

E $13 \cdot 8 \, \mathrm{m \, s^{-2}}$.

5. A $2 \cdot 0$ kg trolley travels in a straight line towards a stationary $5 \cdot 0$ kg trolley as shown.

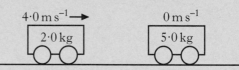

The trolleys collide. After the collision the trolleys move as shown below.

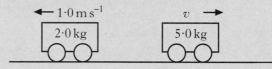

What is the speed v of the $5 \cdot 0$ kg trolley after the collision?

A $0 \cdot 4 \, \mathrm{m \, s^{-1}}$

B $1 \cdot 2 \, \mathrm{m \, s^{-1}}$

C $2 \cdot 0 \, \mathrm{m \, s^{-1}}$

D $2 \cdot 2 \, \mathrm{m \, s^{-1}}$

E $3 \cdot 0 \, \mathrm{m \, s^{-1}}$

6. The density of the gas in a container is initially $5 \cdot 0 \, \mathrm{kg \, m^{-3}}$.

Which of the following increases the density of the gas?

I   Raising the temperature of the gas without changing its mass or volume.

II   Increasing the mass of the gas without changing its volume or temperature.

III Increasing the volume of the gas without changing its mass or temperature.

A   II only

B   III only

C   I and II only

D   II and III only

E   I, II and III

[Turn over

7. For a fixed mass of gas at constant volume

A   the pressure is directly proportional to temperature in °C

B   the pressure is inversely proportional to temperature in °C

C   the pressure is directly proportional to temperature in K

D   the pressure is inversely proportional to temperature in K

E   (pressure × temperature in K) is constant.

8. A potential difference, V, is applied between two metal plates. The plates are 0·15 m apart. A charge of +4·0 mC is released from rest at the positively charged plate as shown.

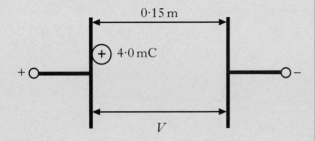

The kinetic energy of the charge just before it hits the negative plate is 8·0 J.

The potential difference between the plates is

A $3 \cdot 2 \times 10^{-2}$ V

B   1·2 V

C   2·0 V

D $2 \cdot 0 \times 10^{3}$ V

E $4 \cdot 0 \times 10^{3}$ V.

9. A battery of e.m.f. 24 V and negligible internal resistance is connected as shown.

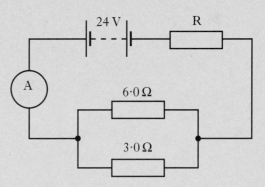

The reading on the ammeter is 2·0 A.

The resistance of R is

A   3·0 Ω

B   4·0 Ω

C   10 Ω

D   12 Ω

E   18 Ω.

10. The diagram shows a Wheatstone Bridge.

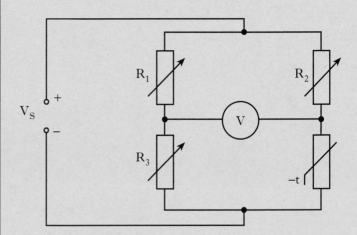

The bridge is initially balanced.

The thermistor is then heated and its resistance decreases. The bridge could be returned to balance by

A   decreasing R_1

B   decreasing R_2

C   increasing R_2

D   increasing R_3

E   increasing V_S.

11. A $25 \cdot 0\,\mu\text{F}$ capacitor is charged until the potential difference across it is $500\,\text{V}$.

The charge stored in the capacitor is

A $5 \cdot 00 \times 10^{-8}\,\text{C}$

B $2 \cdot 00 \times 10^{-5}\,\text{C}$

C $1 \cdot 25 \times 10^{-2}\,\text{C}$

D $1 \cdot 25 \times 10^{4}\,\text{C}$

E $2 \cdot 00 \times 10^{7}\,\text{C}$.

12. A student connects an a.c. supply to an a.c. ammeter and a component **X**.

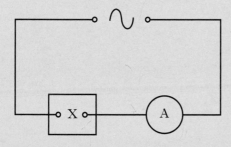

As the frequency of the a.c. supply is steadily increased, the ammeter reading also increases.

Component **X** is a

A capacitor

B diode

C lamp

D resistor

E transistor.

13. An amplifier circuit is set up as shown.

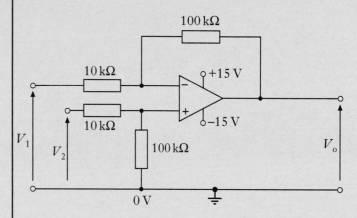

When $V_{\text{o}} = 0 \cdot 60\,\text{V}$ and $V_{1} = 2 \cdot 70\,\text{V}$, what is the value of V_{2}?

A $2 \cdot 10\,\text{V}$

B $2 \cdot 64\,\text{V}$

C $2 \cdot 76\,\text{V}$

D $3 \cdot 30\,\text{V}$

E $8 \cdot 70\,\text{V}$

[Turn over

14. A prism is used to produce a spectrum from a source of white light as shown.

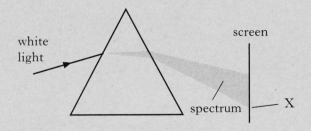

The colour observed at X is noted.

The prism is then replaced by a grating to produce spectra as shown.

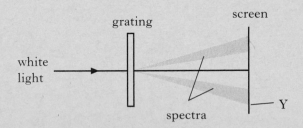

The colour observed at Y is noted.

Which row in the table gives the colour and wavelength of the light observed at X and the light observed at Y?

	Colour of light at X	Wavelength of light at X/nm	Colour of light at Y	Wavelength of light at Y/nm
A	Red	450	Red	450
B	Blue	450	Blue	450
C	Blue	650	Red	450
D	Blue	450	Red	650
E	Red	650	Blue	450

15. A ray of monochromatic light passes into a glass block as shown.

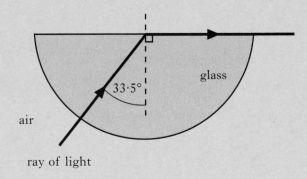

ray of light

The refractive index of the glass for this light is

A   0·03

B   0·55

C   0·87

D   1·20

E   1·81.

16. Which of the following statements about the characteristics of laser light is/are true?

I   It is monochromatic since all the photons have the same frequency.

II   It is coherent because all the photons are in phase.

III   Its irradiance is inversely proportional to the square of the distance from the source.

A   I only

B   I and II only

C   I and III only

D   II and III only

E   I, II and III

17. A student writes the following statements about p-type semiconductor material.

    I Most charge carriers are positive.

    II The p-type material has a positive charge.

    III Impurity atoms in the material have 3 outer electrons.

Which of these statements is/are true?

A   I only

B   II only

C   I and II only

D   I and III only

E   I, II and III

18. A p-n junction diode is forward biased.

Positive and negative charge carriers recombine in the junction region. This causes the emission of

A   a hole

B   an electron

C   an electron-hole pair

D   a proton

E   a photon.

19. A sample of radioactive material has a mass of 20 g. There are 48 000 nuclear decays every minute in this sample.

The activity of the sample is

A   800 Bq

B   2400 Bq

C   48 000 Bq

D   2 400 000 Bq

E   2 880 000 Bq.

20. A sample of body tissue is irradiated by two different types of radiation, X and Y.

The table gives the radiation weighting factor and absorbed dose for each radiation.

Type of radiation	Radiation weighting factor	Absorbed dose/μGy
X	10	5
Y	5	2

The total equivalent dose received by the tissue is

A $0.9\,\mu\text{Sv}$

B $4.5\,\mu\text{Sv}$

C $7.0\,\mu\text{Sv}$

D $40.0\,\mu\text{Sv}$

E $60.0\,\mu\text{Sv}$.

[Turn over

SECTION B

Write your answers to questions 21 to 30 in the answer book.

Marks

21. A basketball player throws a ball with an initial velocity of $6.5 \, \text{m s}^{-1}$ at an angle of $50°$ to the horizontal. The ball is $2.3 \, \text{m}$ above the ground when released.

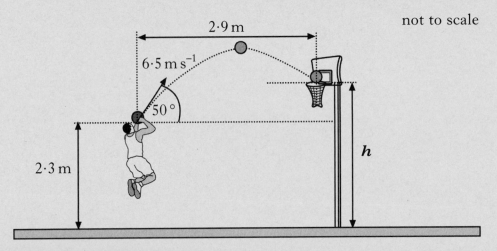

The ball travels a horizontal distance of $2.9 \, \text{m}$ to reach the top of the basket. The effects of air resistance can be ignored.

(a) Calculate:

　(i) the horizontal component of the initial velocity of the ball;　　1

　(ii) the vertical component of the initial velocity of the ball.　　1

(b) Show that the time taken for the ball to reach the basket is $0.69 \, \text{s}$.　　1

(c) Calculate the height h of the top of the basket.　　2

(d) A student observing the player makes the following statement.

"The player should throw the ball with a higher speed at the same angle. The ball would then land in the basket as before but it would take a shorter time to travel the 2.9 metres."

Explain why the student's statement is incorrect.　　2

(7)

Marks

22. Golf clubs are tested to ensure they meet certain standards.

(*a*) In one test, a securely held clubhead is hit by a small steel pendulum. The time of contact between the clubhead and the pendulum is recorded.

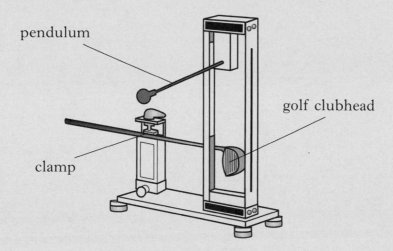

The experiment is repeated several times.

The results are shown.

248 µs   259 µs   251 µs   263 µs   254 µs

(i) Calculate:

(A) the mean contact time between the clubhead and the pendulum;   **1**

(B) the approximate absolute random uncertainty in this value.   **1**

(ii) In this test, the standard required is that the maximum value of the mean contact time must not be greater than 257 µs.

Does the club meet this standard?

You must justify your answer.   **1**

(*b*) In another test, a machine uses a club to hit a stationary golf ball.

The mass of the ball is $4 \cdot 5 \times 10^{-2}$ kg. The ball leaves the club with a speed of $50 \cdot 0$ m s^{-1}. The time of contact between the club and ball is 450 µs.

(i) Calculate the average force exerted on the ball by the club.   **2**

(ii) The test is repeated using a different club and an identical ball. The machine applies the same average force on the ball but with a longer contact time.

What effect, if any, does this have on the speed of the ball as it leaves the club?

Justify your answer.   **2**

(7)

Marks

23. A student is training to become a diver.

(a) The student carries out an experiment to investigate the relationship between the pressure and volume of a fixed mass of gas using the apparatus shown.

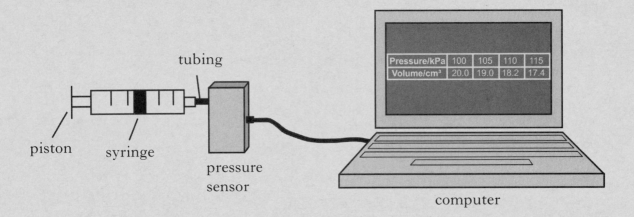

The pressure of the gas is recorded using a pressure sensor connected to a computer. The volume of the gas is also recorded. The student pushes the piston to alter the volume and a series of readings is taken.

The temperature of the gas is constant during the experiment.

The results are shown.

Pressure/kPa	100	105	110	115
Volume/cm^3	20·0	19·0	18·2	17·4

(i) Using **all** the data, establish the relationship between the pressure and volume of the gas. — 2

(ii) Use the kinetic model to explain the change in pressure as the volume of gas decreases. — 2

(b) (i) The density of water in a loch is $1·02 \times 10^3 \, kg \, m^{-3}$. Atmospheric pressure is $1·01 \times 10^5 \, Pa$.

Show that the **total** pressure at a depth of 12·0 m in this loch is $2·21 \times 10^5 \, Pa$. — 2

(ii) At the surface of the loch, the student breathes in a volume of $1·50 \times 10^{-3} \, m^3$ of air.

Calculate the volume this air would occupy at a depth of 12·0 m. The mass and temperature of the air are constant. — 2

(c) At a depth of 12·0 m, the diver fills her lungs with air from her breathing apparatus. She then swims to the surface.

Explain why it would be dangerous for her to hold her breath while doing this. — 2

(10)

Marks

24. A battery of e.m.f. 6·0 V and internal resistance, *r*, is connected to a variable resistor R as shown.

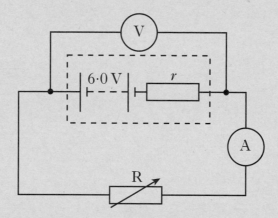

The graph shows how the current in the circuit changes as the resistance of R increases.

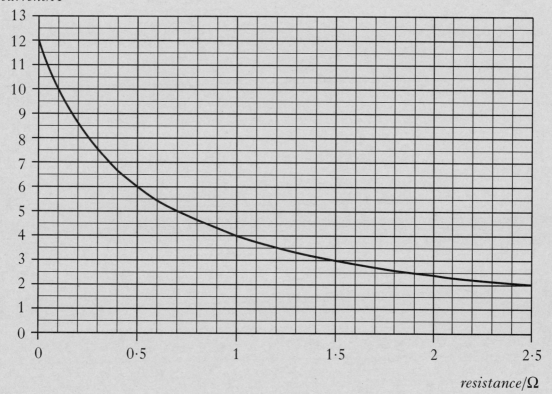

(a) Use information from the graph to calculate:

(i) the lost volts in the circuit when the resistance of R is 1·5 Ω;   **2**

(ii) the internal resistance, *r*, of the battery.   **2**

(b) The resistance of R is now increased.

What effect, if any, does this have on the lost volts?

You must justify your answer.   **2**

(6)

Marks

25. (*a*) A microphone is connected to the input terminals of an oscilloscope.
A tuning fork is made to vibrate and held close to the microphone as shown.

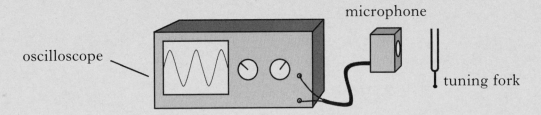

The following diagram shows the trace obtained and the settings on the oscilloscope.

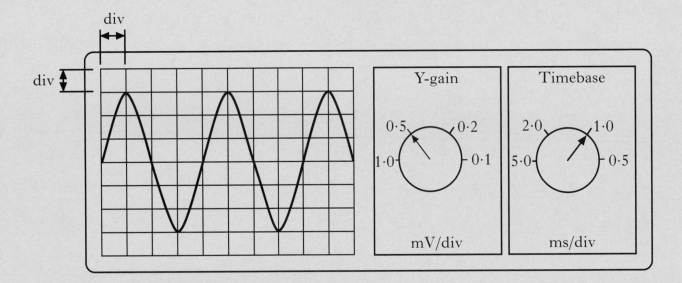

Calculate:

(i) the peak voltage of the signal;   1

(ii) the frequency of the signal.   2

Marks

25.   (continued)

(*b*) To amplify the signal from the microphone, it is connected to an op-amp circuit. The oscilloscope is now connected to the output of the amplifier as shown.

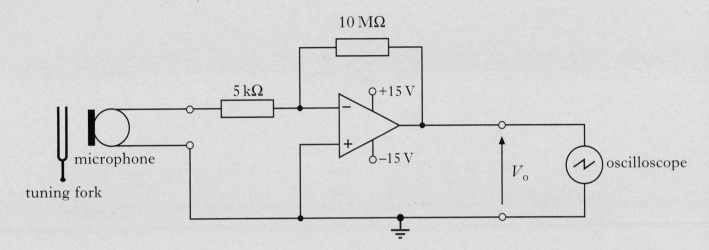

The settings of the oscilloscope are adjusted to show a trace of the amplified signal.

(i)   In which mode is this op-amp being used?   1

(ii)   The peak voltage from the microphone is now 6·2 mV.

Calculate the **r.m.s.** value of the output voltage, V_o, of the op-amp.   3

(iii) With the same input signal and settings on the oscilloscope, the supply voltage to the op-amp is now reduced from ± 15 V to ± 9 V.

What effect does this change have on the trace on the oscilloscope?

Justify your answer.   2

                                                                                        (9)

[Turn over

Marks

26.　A 12 volt battery of negligible internal resistance is connected in a circuit as shown.

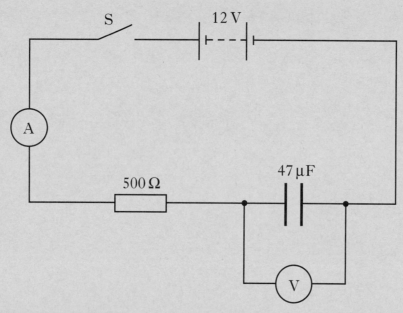

The capacitor is initially uncharged. Switch S is then closed and the capacitor starts to charge.

(a) Sketch a graph of the current against time from the instant switch S is closed. Numerical values are not required.　　1

(b) At one instant during the charging of the capacitor the reading on the ammeter is 5·0 mA.

Calculate the reading on the voltmeter at this instant.　　3

(c) Calculate the **maximum** energy stored in the capacitor in this circuit.　　2

(d) The 500 Ω resistor is now replaced with a 2·0 kΩ resistor.

What effect, if any, does this have on the maximum energy stored in the capacitor?

Justify your answer.　　2

(8)

Marks

27. A laser produces a narrow beam of monochromatic light.

(*a*) Red light from a laser passes through a grating as shown.

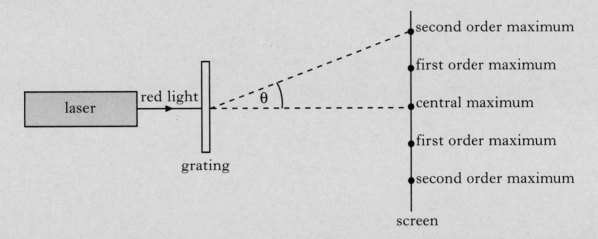

A series of maxima and minima is observed.

Explain in terms of waves how a **minimum** is produced.   1

(*b*) The laser is now replaced by a second laser, which emits blue light.

Explain why the observed maxima are now closer together.   1

(*c*) The wavelength of the blue light from the second laser is 4.73×10^{-7} m. The spacing between the lines on the grating is 2.00×10^{-6} m.

Calculate the angle between the central maximum and the second order maximum.   2

(4)

[Turn over

Marks

28. (*a*) Electrons which orbit the nucleus of an atom can be considered as occupying discrete energy levels.

The following diagram shows some of the energy levels for a particular atom.

E_3 ———————— $-5\cdot2 \times 10^{-19}$ J

E_2 ———————— $-9\cdot0 \times 10^{-19}$ J

E_1 ———————— $-16\cdot2 \times 10^{-19}$ J

E_0 ———————— $-24\cdot6 \times 10^{-19}$ J

(i) Radiation is produced when electrons make transitions from a higher to a lower energy level.

Which transition, between these energy levels, produces radiation with the shortest wavelength?

Justify your answer.   2

(ii) An electron is excited from energy level E_2 to E_3 by absorbing light energy.

What frequency of light is used to excite this electron?   2

(*b*) Another source of light has a frequency of $4\cdot6 \times 10^{14}$ Hz in air.

A ray of this light is directed into a block of transparent material as shown.

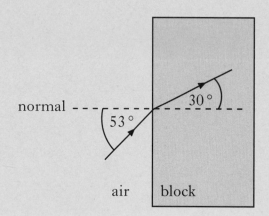

Calculate the wavelength of the light in the block.   3

(7)

Marks

29. Ultraviolet radiation from a lamp is incident on the surface of a metal.

This causes the release of electrons from the surface of the metal.

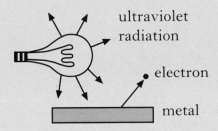

The energy of each photon of ultraviolet light is $5 \cdot 23 \times 10^{-19}$ J.

The work function of the metal is $2 \cdot 56 \times 10^{-19}$ J.

(a) Calculate:

(i) the maximum kinetic energy of an electron released from this metal by this radiation; 1

(ii) the maximum speed of an emitted electron. 2

(b) The source of ultraviolet radiation is now moved further away from the surface of the metal.

State the effect, if any, this has on the maximum speed of an emitted electron.

Justify your answer. 2

(5)

[Turn over

Marks

30. (*a*) Some power stations use nuclear fission reactions to provide energy for generating electricity. The following statement represents a fission reaction.

$$^{235}_{92}U + {}^{1}_{0}n \rightarrow {}^{139}_{57}La + {}^{r}_{42}Mo + 2{}^{1}_{0}n + s{}^{0}_{-1}e$$

(i) Determine the numbers represented by the letters *r* and *s* in the above statement. **1**

(ii) Explain why a nuclear fission reaction releases energy. **1**

(iii) The masses of the particles involved in the reaction are shown in the table.

Particle	Mass/kg
$^{235}_{92}U$	$390 \cdot 173 \times 10^{-27}$
$^{139}_{57}La$	$230 \cdot 584 \times 10^{-27}$
$^{r}_{42}Mo$	$157 \cdot 544 \times 10^{-27}$
$^{1}_{0}n$	$1 \cdot 675 \times 10^{-27}$
$^{0}_{-1}e$	negligible

Calculate the energy released in this reaction. **3**

Marks

30. (continued)

(*b*) One method of reducing the radiation received by a person is by using lead shielding.

In an investigation of the absorption of gamma radiation by lead, the following graph of corrected count rate against thickness of lead is obtained.

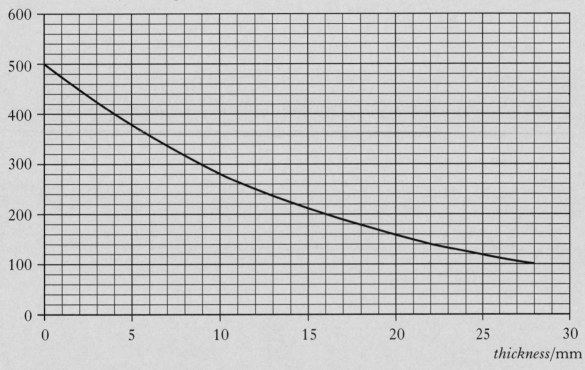

corrected count rate/counts per minute

(i) Determine the half-value thickness of lead for this radiation.   **1**

(ii) With no shielding, the equivalent dose rate a short distance from this source is $200\,\mu\text{Sv}\,\text{h}^{-1}$.

When the source is stored in a lead container, the equivalent dose rate at the same distance falls to $50\,\mu\text{Sv}\,\text{h}^{-1}$.

Calculate the thickness of the lead container.   **1**

(7)

[END OF QUESTION PAPER]

Acknowledgements

Permission has been sought from all relevant copyright holders and Bright Red Publishing is grateful for the use of the following:
A picture of electronically heated gloves. Reproduced with permission of Zanier Sport (2008 page 15).